A RUNWAY IN THE SKY

For Barry with best wishes

Doug Taylor

A RUNWAY IN THE SKY

DOUG TAYLOR

For my family

Published by Doug Taylor

Designed by Paul Barrett Book Production, www.pbbp.co.uk
Edited by Cambridge Editorial Ltd, www.camedit.com
Produced in association with Cambridge Book Group, www.cambookgroup.com
Printed by Short Run Press Ltd, Exeter

ISBN 978-1-5262-0006-8

Contents

About the author

Doug Taylor has had a varied engineering career. He joined the Royal Navy as an Ordnance Artificer Apprentice in 1945 when the Royal Navy still used big gun battleships. He became a marine engineer in the days of steam and then specialised in aircraft engineering, spanning aeroplanes from the last Seafires to the Sea Harrier.

His long struggle to interest the Navy in his Ski Jump invention was a dominant feature of his last years in the service.

After leaving the navy he joined Marconi Avionics, despite being a 'rude mechanical' engineer, where he enjoyed himself designing a variety of weird objects including unmanned aircraft and unmanned mine-hunting submersibles.

He has been awarded the MBE and the Gold Medal for Flight Safety of the Guild of Air Pilots and Navigators.

Note on illustrations

The photographs reproduced in this book have been in the author's possession for many years and it has not been possible to trace and acknowledge all the original copyright holders. The author would be grateful for any information enabling him to correct this omission in future editions of the book.

1

Introduction

It had taken half the morning, or so it seemed, to get me dressed for this occasion and now, trussed like a turkey in flying overalls and unfamiliar G suit, head helmeted and oxygen masked, the time had come. I had been thoroughly briefed and I knew what I had to do. It wasn't very much.

I was sitting in the back seat of the Harrier; we had started up and taxied out and now we were lined up. The ramp seemed very close, like a grey steel wall due to the foreshortening effect. Flight Lieutenant Ellis spoke:

'OK, Doug?'

'Fine, thanks.'

'Off we go, then.'

The engine note rose as the throttle was slammed open, brakes off and the Harrier started to move. This was the moment. We were about to test one of my inventions.

This is not a heroic or glamorous story. It covers long periods of tedium and helpless frustration – just like most people's working lives, in fact. However, I have left most of that out in the hope that what remains will be of more than usual interest.

I wouldn't describe myself as an inventor because I don't think it's a thing you can be, at least not in the same sense as being a doctor or a lawyer or a bricklayer. I call myself an engineer and if asked what sort of engineer I would say that I am of the mechanical and aeronautical persuasion. All the same it has to be admitted that I do invent from time to time but it is largely an involuntary process and not necessarily connected with my profession. I should perhaps add that, once started, I find it difficult to suppress the process. This is not necessarily an advantage, especially in a social context.

Whatever you call yourself, when you have invented once or twice people start thinking of you as an inventor and as such you are perceived in accordance with various stereotypes. Madness and a tendency to wear odd socks, or even no socks, and other unflattering characteristics are attributed to you but they are bearable and can allow the recipient to indulge himself in some harmless amusement by complying with such of these oddities as take his fancy from time to time.

It gets serious, though, when you are expected to invent on demand and to a deadline. Which is, of course what all forms of management want.

If you have this tendency to invent and you work in a large and hierarchical organisation you very soon learn to keep quiet about it, to do it as a kind of secret vice for your own satisfaction, rather as some people are addicted to crosswords. I spent 34 years in the Royal Navy and learned the lesson at an early stage but once or twice I got carried away sufficiently to forget my principles and alert my superiors to the immense benefits I was about to bestow upon the Service. That was rarely well received. However, once, just once, after a lot of time and effort and, above all, with good luck, one of my inventions was crowned with success. Good luck is the vital ingredient.

I don't believe that the Navy is any worse than most organisations when it comes to innovation but it does like things to be properly arranged in a tidy and seamanlike manner.

Government departments and large companies are the same; they are run by methodical people with tidy desks and tidy minds. They are all in favour (they say) of innovation, the lifeblood of industry, etc., etc., but of course it has to be properly planned and managed. Herein lies a problem, because invention is rarely the product of a tidy mind. When most people think of innovation they envisage something like their existing equipment but bigger and better. The better mousetrap must still be easily recognisable as a mousetrap.

Imagine then a modern, highly technical government department such as the Ministry of Defence intent on replacing a piece of equipment with the latest, most innovative model. First they say what they want and when they want it – the staff requirement. Then they beg, bully and persuade the Treasury to allocate the money and place it in the long-term costings while fighting off all the other departments and politicians who think the money could be better spent. Then at last there's an agreement to go ahead. They invite various companies to compete for the job and years later it actually starts.

Progress is good, or more likely not, nevertheless progress occurs. Then suddenly, out of the blue, a mad inventor pops up and claims that he has invented a device that is not only better than your new bit of equipment, but makes it quite unnecessary. The fellow's mad, of course; best to ignore him. If he persists just point out that apart from being mad he is also a person of no consequence with no expertise in this particular field of activity. If that doesn't work then you'll have to go to the trouble of getting your research department to rubbish whatever puny figures the madman has concocted because the programme has been planned, funded and is now being managed and must not be disturbed. The reaction is understandable; one can sympathise. Up to a point.

Inventors do not ask their customers what they want; there's no point. As Henry Ford said, 'If I had asked my customers what they wanted they would have said, 'A faster horse'.'

As a general rule it is probably fair to say that no one wants to be thought of as a narrow-minded obstacle to innovation but few find it easy to accept the genuinely new and unfamiliar, except in fashion, which is merely variations on a theme. Inventions are most readily accepted, albeit reluctantly, when they meet a long-felt but seemingly impossible attainment or need or create an entirely new and welcome capability. It is folly to put significant money or effort into any invention that does not match these criteria.

The budding inventor usually becomes aware of all this at a very early stage and, because he can't stop inventing, gets on with life and treats inventing as a hobby.

It follows that most inventions never see the light of day, even those that meet the criteria. To the inventor, or at least to this inventor, that doesn't matter too much generally. The satisfaction comes from the triumphant solution to the problem – the completion of a particularly demanding crossword may be an appropriate analogy. The dreary, laborious and boring business of getting someone to provide the wherewithal to turn it into functioning hardware usually has no appeal. So another criterion emerges: if you really seriously want to turn your idea into reality it should be sufficiently simple for you to accomplish the feat unaided in your garden shed. Some courageous people have done exactly that, including the inventor of what I consider to be one of the outstanding, if unspectacular, inventions of the 20th century, namely the Workmate. That meets all the criteria but nonetheless the man who invented it had enormous difficulties in getting it into production.

No sane person tries to earn his living by inventing.

Apart from the difficulties inherent in the business of invention there exist other, frequently more formidable obstacles to the inventor. They can be classified as social and political and will depend on the peculiar circumstances of each individual. In my case I was an undistinguished member of the Fleet Air Arm of the Royal Navy and an engineer. Being an engineer is, or was, a social matter. The Navy has had difficulty in adjusting to the recent (19th-century) addition of engineers to its ranks. This is in part a reflection of British snobbery and its aversion to any activity that involves getting your hands dirty.

And it has to be admitted that in my day we engineers often had dirty hands and wore oily overalls to boot. Such people were only reluctantly tolerated and then only when not displaying the grime of their trade. Indeed, I have known engineer officers who have advanced as a consequence of a studious avoidance of ever appearing in overalls.

The Fleet Air Arm (FAA) is an even more recent arrival, with social implications on the naval scene, as there has long been dispute about whether aviation, apart from being very dangerous, was really a gentlemanly occupation, requiring, as it does, a degree of technical knowledge. However, nowadays aviators are generally acceptable, perhaps because they wear clean overalls. Aviators are generally rather more tolerant of engineers, provided they know their place and do not venture their opinions on subjects outside their specialisation.

The political obstacles to my activities arose entirely from the existence of the Royal Air Force and its insistence that it alone should provide the air defence for the UK and its forces, despite notable failures to do so, particularly at sea. The RAF therefore pursues a consistent policy of attempting to absorb and thereby eliminate the Fleet Air Arm. It has done this with considerable success over the years, and so done almost as much damage to our country's defences as have our politicians. As a result the FAA has to fight for its existence in the eyes of politicians and public who know little and care less of the FAA's achievements. The Navy was therefore understandably nervous about any of my activities that might come to light and provide ammunition to the RAF enemy.

With all of the above in mind it is tempting to wonder if my invention was of such truly remarkable virtues that it was bound to survive and come to fruition. I'm very proud of it but I would not claim such exceptional qualities for it. It very nearly fell by the wayside several times, sometimes when I got fed up with bashing my head against the obduracy and ignorance of the opposition and other times when it appeared to be overtaken by events. It survived as a result of persistence and luck.

Especially luck.

2

Some Background Scenery

At the age of 11 I invented the opposed piston two-stroke engine. I intended to use this precocious achievement to make my fortune when I grew up. Imagine then my disappointment when shortly thereafter I discovered that not only had someone already invented it but that it was powering some of the German aircraft that were dropping bombs in my direction. This was my first experience of what was to become a fairly common occurrence. You invent something of extraordinary brilliance only to discover that someone else did it years ago.

Like many boys of my age I took a keen interest in the air activity of which there was no lack in 1940 and 1941. I took a magazine called *The Aeroplane Spotter* (price 3d weekly) and so was able not only to recognise virtually every aeroplane in the sky but could also tell you what mark it was and (almost) its modification state. Since many of the aircraft on view at any given moment were liable to be German, this ability undoubtedly had survival value. My father, who was a sergeant in an anti-aircraft battery at the time, was home on leave on one occasion during an air raid and cheered wildly as one of the aircraft came howling vertically downward to end in a fireball about half a mile away. I pointed out that his cheers were misplaced as the aircraft in question was a Spitfire. I then compounded my offence by venturing an opinion on the standards of aircraft recognition in the Royal Artillery. The ensuing clip round the ear hole (for being right) also had survival value.

Despite my keen interest in aircraft, my first love was boats – boats of almost any description but principally canoes. I had a book that had a chapter on 'How to build a canvas canoe for only £5'. Only £5! It was a sum beyond any rational probability of being available to me but I knew that chapter by heart – after all you never knew how things might turn out. Second best in my estimation would be a Pram dinghy but that was even further beyond any means of mine and so not worth serious consideration. As it happened I did build the Pram dinghy 20 years or so later and have it still. At that time my only experience of boats was the Woolwich free ferry on which I spent a great deal of my spare time. Although each voyage was only half a mile across the river there was much shipping to be seen going and coming from the London docks. But the icing on the cake was the ferries' engine rooms, which were open to view and exuded a delightful aroma of steam and hot oil from the gleaming creations of polished steel and brass moving so majestically and quietly below. This probably awoke in me an interest in engineering but busy screeching turbines have never had, for me, the magic appeal of those old engines that, even so long ago, were totally obsolete.

It was wartime and so of course the armed services were in the forefront of events. For me at 12 or 13 there were two career choices because both were boats writ large, either the Royal Navy or the Merchant Service, and I really had no strong preference. It was just a matter of which would be first to offer an opportunity. I had not long to wait. I saw, in some publication or other, a notice to the effect that boys wishing to become officer cadets at the Royal Naval College Dartmouth should apply forthwith to take the entrance examination and should have parental consent and a letter of recommendation from their headmaster. This, I thought, would suit me nicely.

Parental consent was no serious problem, my father was overseas with the Army so consent was in my mother's hands and I knew she could be persuaded. I wasn't so sure about the headmaster but it seemed to me that he had no reason not to recommend me, so I duly requested an audience. His response was unexpected. He said that it would be a waste of my time and his as there was no chance that I would be accepted. I found this difficult to believe. Having looked at the exam requirements I reckoned that I had a good chance of passing. He agreed, but gently made me aware of some facts of life at that time. He pointed out that there was also an interview and that this was the main hurdle, regardless of exam results. The sad fact was, he said, that in his experience the likelihood of a grammar school boy like me being admitted to Dartmouth was almost nonexistent. The best thing to do was to forget Dartmouth, wait a year or two, and think again. Very reluctantly I accepted his advice, which, with hindsight, I know was sound. As things turned out, five years later I was admitted to Dartmouth in rather different circumstances.

Another interest of mine was guns – an interest shared by most of my friends. This interest was entirely innocent and not to be compared with that of contemporary inner city youths. Also, though we were surrounded by weapons of various sizes, guns were not easy to obtain other than for waging war. Fortunately the absent father of one of my friends had a workshop that included a lathe, so we made our own. We started with miniature cannon but soon progressed to pistols and in this endeavour we hit a problem that will be familiar to many generations of boys, which is that it is very difficult to make gunpowder, or rather difficult to make gunpowder that will explode. When ignited, ours would invariably fizz and bubble but do nothing more interesting, not even when thoroughly dried in some mother's oven. Luckily the Luftwaffe came unwittingly to solve the problem. One night after a period of ground-softening heavy rain the Luftwaffe unloaded a container of incendiary bombs across a neighbouring golf course where they stuck, unexploded, like some sparse crop to be harvested by me and my friends. We gathered about half a dozen and unscrewed the cast iron nose caps that contained the fuse that should have set them off. We could then tip out the thermite powder that started the bomb burning. Next, to the tail. Knock off the fin and, if there's an anti-personnel detonator under it, remove same. What's left is the solid magnesium carcass of the bomb. This we would set up between centres in the lathe. In the tool holder we put a thread chaser – as fine a thread as we could find – mounted at a slight angle. Then, with a small cut and a fine feed the lathe would produce masses of very fine magnesium swarf to be pounded up into fine magnesium flakes. The only other ingredient required was permanganate of potash, sold by all chemists for a variety of innocent uses and so not likely to provoke questions. A 50/50 mixture of these ingredients would explode readily, even when slightly damp, so we had ample powder for our pistols. The drawbacks were barrel fouling and a muzzle flash that lit up the countryside when fired at night. A further advantage was that the addition of sulphur to the mixture slowed the burn rate so that the mixture made a very amenable rocket fuel. I have no guilt about revealing these lethal secrets because one of the ingredients is now, mercifully, no longer available and because any boy wishing to do the same thing nowadays has access to the Internet and a wealth of more sophisticated information.

My interest in boats and the sea, together with my interest in guns, weighted my inclination towards the Royal Navy rather than the Merchant Service, admirable as the latter was, and in a year or two there came another notice of an entrance examination for boys wishing to become Artificer Apprentices in the Royal Navy. There was no interview and it was clear that grammar school boys were welcome to apply, so I did. I took the exam and expressed a wish to be an Ordnance Artificer Apprentice. (Other alternatives were Engine Room, Electrical and Aircraft.) In due course the exam results were announced. I had passed and I was instructed to report to the Royal Naval Artificer's Training Establishment at Torpoint in Cornwall on 1 August 1945.

On the day my dad came up to London with me. He was home on leave, awaiting demobilisation and by no means happy about my joining the Navy. Despite his misgivings he saw me off on the 10.30 from Paddington. This train had the rather grandiose name of the Cornish Riviera Express. To me it had all the mystery and excitement of a camel train to Samarkand; it was going into unknown territory, far away in the west. Cornwall! I had never been further west than Brighton so the next four hours were an adventure in themselves.

When the train arrived at Plymouth we were met by a Petty Officer and rounded up into a Naval lorry for the short journey across the river Tamar to Torpoint. We crossed the river on the chain ferry and a splendid sight met our eyes because the Hamoase was full of warships. In 1945 the Royal Navy was bigger than any other navy in the world except that of the USA and it showed, magnificently. The nearest ships to the ferry were the battle cruisers Renown, Ramilles and Resolution, moored side by side. With their 15-inch guns the trio were an impressive pile of steel armour. We were not to know then that they were used mainly for accommodation and practical training and already earmarked for the scrapheap.

Once across the river it was only two miles or so to our destination, HMS Fisgard – a hutted shore establishment built on the side of a Cornish hill. It's all gone now but it was to be our home for the next four years – or so we thought. Now at the age of 16, four years is an eternity and I wanted to go to sea as soon as possible. I had managed to reconcile this conflict in my mind simply by believing that during four years in the Navy it was quite impossible that we should not go to sea in one way or another.

The immediate present was more mundane but interesting enough because everything was new. We collected our kit and our poor-quality, ill-fitting uniforms and learned the rudiments of drill on the parade ground. In time the parade ground and the activities upon it would become the most hated parts of our new life. It was on the parade ground that the first signs of dissent appeared – and were swiftly quashed. One apprentice disputed an order on the grounds that it was unfair and anyway the Navy couldn't make him do it. The Petty Officer was, considering the circumstances, remarkably gentle: 'My son,' he said, 'the Navy can do anything it likes with you' – pause – 'except make you pregnant.'

After three weeks of this we were all, quite unexpectedly, sent home on a long weekend's leave. The war against Japan had come dramatically to an end with the dropping of the atomic bombs.

For our part in this war, all three weeks of it, we were awarded a medal that I still, very rarely, have occasion to wear.

On our return things began to get more earnest and our life began to revolve around 'school' and 'factory' rather than the parade ground. School consisted of science, physics, maths and engineering subjects together with some highly biased naval history centred on the sacred figure of Lord Nelson. We became thoroughly fed up with Nelson. School happened on two mornings and three evenings a week and the standard required was high even for those days. Today I would guess that it approximated to A-level. All the rest was factory and the standard required was the best. My trade was to be, like most, that of fitter and turner.

We were introduced to our trade with the issue of a tool box containing a large hammer, a smaller hammer, two cold chisels and two files, one rough and one smooth. With these basic tools to start with we would find that we would be making most of our tool kits in the next few months. For now we were each presented with two ugly-looking pieces of metal. The larger was a square slab of rough cast iron about six inches square and rather more than an inch thick. The smaller was a piece of mild steel bar about two inches in diameter and four inches long. A square outline was marked out on its ends. Our instructor explained that, using our chisels, we would remove the metal almost down to the markings and then smooth it off down to the lines (but no further) using the files. It seemed to us that there was an awful lot of metal to be removed. One voice in particular thought it was all a joke: 'You can't chip steel, Chief,' he said, scornfully.

The Chief showed him the error of his ways. He clamped the steel bar in the vice, seized hammer and chisel and with consummate ease was soon producing spectacular curling slivers of steel. Obviously it wasn't as difficult as we had thought so we all set to with enthusiasm. By the end of the session little enthusiasm remained; everyone had blisters on one hand and cuts and bruises on the other. Some had inflicted wounds requiring the attention of the sick bay. As time went on our hammers hit the target more frequently, metal was removed more efficiently and we got down to the marked line at last. The next exercise taught us how to mark out a new line 1/4 inch below the first. We then chipped and filed our way down to the new line. We were becoming hardened by now and when invited to attack the cast iron we did so with confidence, which, we soon realised, was misplaced because the cast iron was altogether nastier, dirtier and generally difficult.

Weeks later both bits of metal were much smaller and our hands were scarred but hard. Sixty-odd years later my hands are soft again but I can still hit the head of a chisel every time and I can file flat.

Saturday and Sunday afternoons were free, we could go 'ashore' but only by catching the 'liberty boat' at fixed times. This involved falling in in threes by the main gate and being inspected by the Officer of the Watch. Those passing his inspection were then marched out through the gate to enjoy the fleshpots of Torpoint and Plymouth.

Sunday mornings were taken up with divisions and church. Divisions was a parade in best uniforms and shiny boots to be closely inspected by our Divisional Officer followed by marching about to the recorded music of a brass band. Not an interesting experience at the best of times, it was also fraught with the danger that one's appearance would lack some essential brilliance and result in the ordering of a kit muster to take place just as you thought you were going ashore. Even worse, more serious scruffiness could result in your leave being stopped for days at a time by the Divisional Officer – a godlike creature with the rank of lieutenant. We knew that he was godlike because we had heard him addressed by his first name by God himself. God was the Captain. There were, we understood, greater gods but we never saw them. We didn't see a lot of our Divisional Officer, though what we did see seemed decent enough, albeit godlike. Authority manifested itself mainly in the shape of Chief and Petty Officers who were human and in some cases all too easily provoked to wrath. This particularly applied to those of the gunnery or physical training variety. Our factory instructors were mostly retired artificers and these we held in great esteem.

After divisions came church. The Navy treated church and religion very seriously because it was considered to be good for the troops. Atheism was not allowed; if you were uncertain of your spiritual allegiance you were automatically assigned to the Church of England. It followed that Sunday worship was compulsory. After divisions we were marched to the church and halted at the door. Then came the command, 'Roman Catholics and fancy religions fall out,' and the rest of us went inside. Fancy religions meant Jews and nonconformists and they didn't escape if an appropriate place of worship was handy.

This compulsion met with no protest from our padre, quite the reverse. Soon after we joined he asked the class which of us were confirmed. Perhaps two raised their hands. The rest of us had no idea what he meant so we kept quiet in case we accidentally volunteered for something. In that case, he said, we would all attend confirmation classes on one of our free evenings each week. Not one of us protested; it never occurred to us that we could, let alone should, and so the classes went on for weeks. I now have no recollection at all of their content but they eventually resulted in a ceremony in the church in which we were gabbled over, two at a time by the Bishop of Truro and the business was done. The church was a recurrent and unnecessary irritant to me during my time in the Navy and not one of my seeking.

Generally speaking I enjoyed my time as an Artificer Apprentice. We were made to feel that we were important to and valued by the Navy and our work was interesting and fulfilling. In fact I can think of nothing that has given such consistent pleasure and sense of achievement as learning my trade as a fitter and turner and I still take pride in such skills as remain to me.

In my day there was a scheme whereby after two years we were eligible for selection for promotion to Cadet (E) and the fast track to commissioned rank. To be eligible, the rules were that you must have consistently achieved better than 70% in the school and factory exams. I always got about 80% for my practical work but in school I never was able to do better than about 65% so I did not see myself as a contender (maths was my weakness) and accepted the situation. In any case it was a long shot and, as I had seen before, there was an interview, which was the main hurdle. I was therefore very surprised, after two years, to find myself one of five apprentices nominated by the Divisional Officer as candidates for a cadetship. I didn't let myself get too excited by the event; my headmaster's words were still fresh in my mind.

In due course the five of us were taken up to the Admiralty in London for the interview board. By now it *was* serious and we were very nervous. My recollection of the interview is confused. I was stunned by the array of brass bound sleeves confronting me. I do recall being asked what I felt about equality for women. The truthful answer would have been that I had never given it a thought, not being one. I thought that this might seem a little bit unfashionable so I said that I was all in favour. The reaction of one of the Admirals was distinctly hostile: 'So you're opposed to marriage allowance, then,' he snorted. I've no idea how I answered that. I just assumed that I had blown my chances completely.

Back we went to HMS Fisgard but we had not long to wait in suspense. A few days later the five of us were called to the presence of the Captain. He didn't waste time. 'It was a good effort and I warned you that few, if any, would be accepted. The Admiralty only accepted Taylor.'

Two days later I was on my way home.

3

Dartmouth

I was in a dreamlike state for a little while, unable to believe my good fortune but there was much to be done and it was just as well that there was six week's leave to do it in. There were uniforms to be ordered and made and the rest of my kit to be procured. For these purposes we went to Gieves Ltd of Old Bond Street, an establishment hitherto far out of my league or that of my family. My father came with me on these expeditions though I'm sure that he was as out of his depth as I was. Up to this point all my worldly goods had fitted comfortably into my kit bag with plenty of room to spare. Now I had to have two large trunks (one made of steel) and two large suitcases and the kit to fill them had to be bought; it was not a Navy issue. I was given a uniform allowance for this purpose but, as with many financial interactions with the Admiralty, it was not enough. My dear dad made up the difference out of his war gratuity. It's one of the things dads are for. My uniforms were a cause of wonder, the quality of the material was such a contrast to the coarse cheap material that I had been used to. In fact I had not realised how coarse and cheap they were until compared with Gieves superfine. Eventually everything was assembled and the day came for me to return to the Navy but now with a marginally exalted status. Once again I was to catch the 10.30 from Paddington but for a different destination and once again my dad saw me off.

At Paddington the new cadets in gleaming new uniforms and accompanied by kit bearing no scars stood out and we naturally gravitated together in one carriage. For me it was to be an eye-opening experience, my first close contact with the upper middle classes. They all spoke posh, of course, but that was only to be expected. What was so odd was that their conversation seemed to be all about schools. I had not appreciated that, of course, they had all just left their schools and which school they had attended was very important to them. Inevitably I was asked about mine and with some feeling of superiority I explained that I hadn't just left school, I had already been in the Navy for two years. This statement made less impression than I felt it deserved and clearly was a source of puzzlement to one questioner at least. 'And where was your preppah?', he enquired.

At Kingswear we were met by a Petty Officer and escorted to the ferry across the river to Dartmouth and the Royal Naval College standing high over the town. It is a very impressive building and was a far cry from the hutted camp of HMS Fisgard.

Life at Dartmouth was also different. For a start the pace was even more hectic and the atmosphere was oppressive – to me at any rate. My previous two years of naval experience were no help to me; in some ways, in fact, rather the reverse. As before, immediate authority was in the shape of Chief and Petty Officers who were all from the seaman or gunnery branch with a deep dislike of artificers who, they believed, gave themselves unjustified airs. My past, it was made clear, was not an endearing feature.

At the head of immediate authority came the Cadet Gunner, a Warrant Officer (in those days the bearer of one gold ring on his sleeve), and a figure to be feared. We all felt his wrath in some tangible form on a daily basis, with one exception. One cadet was rumoured, I don't know whether correctly or not, to be a member of the aristocracy. On him the Cadet Gunner fawned in the most sickening manner. In later years colleagues have spoken of the Cadet Gunner, not without affection, as a 'character'. I remember him as a bully and a snob and one to be feared.

Our term officer was a Lieutenant Commander, a genial and friendly soul but far too remote to have much obvious impact on our daily lives. Even more remote was the Captain of the college but he made a point of meeting each of us in his office during our first week or two at Dartmouth. Naturally, for the cadets, this interview was a bit of an ordeal but the Captain did his best to put each of us at ease. When it came to my turn I was just beginning to relax a little when the Captain told me that in his opinion officers should not be promoted from the lower deck. He said that this was not due to any shortcomings in ability or intelligence but simply that our background and the attitudes we would inevitably form would make it extremely difficult if not impossible for such a person to adapt adequately to the lifestyle required of a naval officer. This was said kindly and, he pointed out, was a view obviously not shared by the Admiralty, 'so we will both have to make the best of it.' He wished me the best of luck and we never had occasion to speak again.

Naturally I was not at all happy to hear this and it did nothing to help my self-confidence but I am bound to admit that with 60 years' hindsight and in the context of the times, he had a point. The products of Dartmouth were what the Navy wanted and generally they lived up to the Navy's expectations splendidly.

One thing Dartmouth did have was boats – boats in quantity and in considerable diversity, from skiffs and sailing dinghies through whalers and cutters to gigs and pinnaces. There was even an old-type steam pinnace complete with polished brass funnel. There were no canoes but I felt able to live with that omission. We were encouraged to make use of these boats; much of our daily routine was boatwork and I also used them during our very limited leisure time. I soon discovered that boats were earnest matters, not to be confused with fun and that a fender left in position for half a second longer than strictly necessary had disciplinary consequences. Despite this, fun was had but discreetly.

The thing I found oppressive was the attitude of authority, a great contrast to what I had been used to. At Fisgard we were made to feel valued; at Dartmouth it was made clear that we were creatures of no significance whose existence was justified only by our puny contributions to the good of the Royal Navy, which, in all circumstances, came first.

If, at some later time, we were foolish enough to encumber ourselves with wives and families that was our affair but the Navy still came first. Here, for the first time, we heard the story, undoubtedly apocryphal, of the destroyer's First Lieutenant. The destroyer is alongside in harbour but about to sail for exercises when a telegram is received telling the First Lieutenant that his wife/child is very ill. The Captain sends for the First Lieutenant and breaks the news adding, 'Take a week's leave, Number 1, we will manage without you for a while.' To this the paragon replies, 'Thank you, sir, but no. My place is with the ship.'

Those were the days. It was things like this that made Dartmouth hard to take for me and I was very glad when the time came to leave and join the training cruiser. Off to sea at last.

4

The Training Cruiser

In those days the Royal Navy could devote a fairly major warship solely to the training of its junior officers. The warship concerned was the 10,000-ton cruiser Devonshire, originally armed with eight 8-inch guns but the X turret had been removed and replaced with what amounted to steel sheds. These were the cadets' classrooms. Aside from classwork the cadets performed all the ship's duties and tasks from the most menial up to officer of the watch, though overseen sufficiently to avoid disasters. We lived on messdecks and slept in hammocks like the sailors but, in fact, separated from them.

In the classrooms we learned navigation, seamanship, signals, etc., and even a smattering of engineering, all of which we were then able to put into practice on the spot. We soon learned that a junior seaman's tasks included a great many miserable jobs like washing paintwork and polishing brass, though these were not without compensations. I was in the quarterdeck division and on the quarterdeck was a magnificent gunmetal staghorn bollard. From time to time it fell to me to polish this edifice and in order to do the lower parts one had to lie on one's back. On a warm sunny day this could be very agreeable, applying the Brasso in leisurely mode and languidly rubbing it off under the very eyes of authority. This restful pastime could be extended with impunity provided that regular movement was maintained; the danger was of falling asleep.

Our first voyage began in January when we left Devonport bound for the West Indies, arguably the ideal destination for an initiation into seafaring. As the ship got further south the weather became pleasantly warm and the grey Atlantic became impossibly blue. One day the ship was stopped and 'hands to bathe' was piped. It was an eerie experience to swim on top of a mile or more of water depth, to peer down through an incredible clarity and see the screws and rudder in every detail against that background of blue infinity.

A day or two later came another marvel – flying fish exploding out of the water under our bows. And then one morning there appeared a patch of stubble on the horizon that soon grew into palm trees on a beach of golden sand. We had arrived in the West Indies and it was living up to expectations. No subsequent landfall had anything like the magic of this first one.

While we were in the West Indies we had our first experience of the kind of typical minor crisis which the Navy is frequently called upon to resolve. We were in Montego Bay, Jamaica, when a signal came telling us to go at maximum speed to British Honduras because the neighbours, Guatemala, threatened to invade. So off we went and landed a platoon of marines who persuaded the Guatemalans to be good neighbours by peaceful threats. I was never in any ship that didn't have at least one episode like this during the course of a commission.

We also discovered that the sea had its nasty side. A month or two later we were far from palm trees and tropical beaches. It was a dark, unpleasant night and blowing a full gale as we passed Cape Wrath. I was in my hammock and sound asleep when I was woken by a pain in my side and the most appalling, roaring, crashing noise. The pain in my side was due to pressure from the edge of some fan trunking on the deckhead (ceiling). The noise was made by the messdeck tables, forms and lockers crashing down to the lower side, having come out of their sockets. The ship had broached in a following sea and been laid on her beam ends. She seemed to lie on her side for a long time before coming back up. Altogether a very alarming experience.

Later that summer we went to the Baltic and to the lovely Stockholm archipelago – another magical place.

On the last day at sea I began to feel a bit off-colour and didn't fancy any breakfast. This was unusual to say the least but I paid no great heed as we were all looking forward to arriving in Devonport that evening and going on leave the next day. However, in the next hour or two I began to feel a great deal worse, left whatever I was doing and went down to the messdeck. What happened from then on is very blurred in my memory. The pain became intense and it seems that I was found curled up on the deck. In the sick bay I was given a shot of morphine and agony was changed to euphoria. I was blissfully happy and can well understand how people become addicted to the stuff. Later I was told that there was some discussion as to whether or not I should be operated on at sea but it was decided to increase speed and put me ashore at Devonport. There I was strapped into a stretcher and swung over the side on the ship's crane, still burbling happily, and taken to the Royal Naval Hospital at Stonehouse where my appendix was removed via an incision that commands medical admiration to this day. No keyhole surgery in those days.

I woke next day to find a white-coated figure leaning over me and asking how I felt. He spoke beautifully fluting Oxbridge, so I replied to the effect of, 'Better, thank you, sir.' Things went on in this vein for a little while, becoming slightly strained until he said, 'You don't have to call me sir, sir – I'm a sick bay attendant.' Those were the days of National Service and the Navy was very choosy about whom it would condescend to take.

After two weeks in hospital I was sent home on three weeks' sick leave with instructions to lift nothing heavier than a cup of tea. I was also elevated to the rank of Midshipman. After three weeks I was to report to the RN Hospital at Gillingham for a medical check that I was fit for duty.

I was very happy with this arrangement. Three weeks of the blessings of the land which included late rising, home cooking and the company of girls. My dismay can therefore be imagined when, after less than a week of these delights I received a telegram telling me to report on board HMS Ocean in the King George V Dock, Glasgow, in five days' time. After the initial shock came calmer reflection. Obviously the Admiralty did not understand the situation and all that was required was for me to get the RN Hospital at Gillingham to put them right while I continued my life of ease. I duly appeared at the hospital and was seen by the Surgeon Rear Admiral, no less. He examined me and pronounced a totally unacceptable verdict. 'I really shouldn't do this,' he said, 'but I know how eager you young fellows are to get back to sea. But you must avoid climbing ladders.'

I could muster no coherent response to the daft old buffer who clearly had no idea of my aspirations and presumably thought that aircraft carriers were ladderless. As it all sank in I realised that there was much to be done. First of all I must get my uniforms back from the cleaners and get the Midshipman's patches sewn on. This was the point at which Murphy's law manifested itself because when I asked at the cleaners for my uniforms back *now* they said they had been sent to Liverpool for the actual cleaning and could not be retrieved in less than a week. Threats and pleading had no effect. I was in trouble. In desperation I turned to Gieves. They had a 'can do' reputation to live up to. Once again dad came with me (I was stitched together, remember) and we explained my problem. The man came back in a minute or two with a uniform that fitted me perfectly. True, it had a Lieutenant's stripes but these were soon removed and a Midshipman's patches substituted. My relief and gratitude were profound and quite overshadowed the fact that I now had a new uniform that was unplanned and that I could ill afford.

Next day I caught the overnight train to Glasgow.

5

A Midshipman's Life for Me

Glasgow in a drizzly, cold dawn had no charms. I had not slept much and felt a strong urge for a cup of tea and breakfast. Luckily there was no difficulty in finding a taxi to take me to the King George V dock, where a surprise awaited me. It was a dry dock. The ship was there all right but in a dry dock. I hobbled on board and reported to the officer of the watch, as one does, and found a further surprise – they were not expecting me nor did they feel any pressing need of my services. Looking around the quarterdeck I became aware of another anomaly; in the middle there was a hole about 12 feet square and peering down the hole one saw, past two or three decks, the bottom of the dock and the conspicuous absence of a rudder.

Six weeks later we flooded up and moved out to sea complete with rudder. To this day I wonder what imagined urgent requirement necessitated the recall of a midshipman to a ship with a hole in its bottom. And the Navy still owes me two weeks' leave.

I served in a fair number of the Royal Navy's aircraft carriers – Ocean, Indomitable, Glory, Eagle, Ark Royal and Victorious – some of them more than once. My first proper ship, as distinct from the training cruiser and therefore a bit special, was HMS Ocean, a light fleet carrier of the Colossus class. As soon as we got to sea and 'flying stations' was sounded I made my way to the 'goofers' on the island to watch the fun. The first aeroplane to launch was a Firefly. I had never seen a launch from a carrier so I was eager to miss nothing. The Firefly trundled along the flight deck surprisingly slowly, I thought. When it came to the end it began to lose height and hit the sea about 300 yards ahead of the ship. By the time the splash had subsided the crew were already out and seconds later the aeroplane was gone. They were soon picked up by a whaler from our attendant destroyer but there could be no doubt that this was a dangerous business. The pilot's friends appeared to treat the event with an unseemly levity: typical, they said, of H—, the dozy so-and-so had had his propeller in coarse pitch. Later in the day there was another emergency: 'Crash on deck' came the broadcast. I rushed back up to the 'goofers' to see a rather bent Seafire with its nose embedded in the barrier, having missed all the arresting wires. The day's events provoked a certain amount of deep thought because we midshipmen were shortly to have our 'air experience' flights and one of us at least had mixed feelings about the prospect. The day came when I was to have my first ever flight and my name appeared on the flying programme. I was not at all happy to find that my pilot was to be none other than Lieutenant H— who, in my opinion, treated his recent swim and the reason for it far too lightly. Our task was 'fighter affiliation'. I had no idea what that was but the whole event was an exciting and totally confusing mixture of new sensations. I saw nothing but sea, sky and Seafires flashing past like minnows in a stream. There was a seemingly non-stop stream of chatter on the radio, most of which meant nothing to me. At one point I ventured to speak to my pilot who, so far, had not spoken to me. 'Shut up, Mid,' came the reply and we never spoke again. My first flight, 1948.

There's no doubt that the flight deck of an aircraft carrier is a dangerous place and it was particularly so in those days of piston-engined aircraft and straight flight decks. With the aeroplane in landing attitude the pilot's view ahead varied from poor to very poor so he had to be guided by a landing signals officer or 'batsman'. Not surprisingly it was not at all unusual for the aeroplane to miss all the arrestor wires to be stopped by the barrier. This was a violent process that damaged the aeroplane but not, normally, its crew. However, the propeller usually shattered, scattering lethal chunks of wood or metallic shrapnel, according to the material involved. Mercifully fire was relatively rare but when it happened it complicated the situation no end. There were plenty of other perils. Rotating propellers are invisible and lethal, slipstream can blow you overboard, moving arrestor wires break bones or worse and aircraft lifts descend without warning to create an extremely large pitfall. At night these dangers are magnified as no lighting is allowed, apart from dimmed torches, to avoid damaging the pilots' night vision. Add to all this the noise of engines and the buffeting of a 30-knots plus wind and it is easy to become confused or distracted. You get used to it but, as with freedom, the price of survival is eternal vigilance. Nowadays with jet aircraft that give the pilot a good forward view, the angled deck needing no barrier and the projector sight to guide the pilot, things are much better – but the flight deck is still no place for the unwary.

At this stage we midshipmen had no part in these flight deck operations. We were continuing our training at a working level and undertaking more responsible tasks. We soon realised that aircraft carrier operations were best undertaken well away from land and other shipping and that the working day began well before dawn and ended, or rather paused, at around midnight. Not for us the leisurely cruise among lovely Greek islands with shore leave in some exotic holiday destination. It was thundering up and down an empty stretch of ocean with an occasional week-end in Grand Harbour, Valleta, if we were lucky or, more often, the limited social scene in Marsaxlokk Bay. However, our training included a few weeks in small ships, in my case the destroyer Troubridge, which had had an exceptionally busy war and was showing her age. One day I was in the engine room supervising some stokers scraping old paint from the forward bulkhead, when one of their scrapers went clean through the bulkhead into the boiler room. A well-meant but misguided attempt to scrape back to sound metal soon resulted in a hole big enough for head and shoulders. I could never feel too confident about that ship's watertight integrity but it was a pleasant interlude, even though we never did get to the Greek islands.

As a midshipman at sea I always felt a bit in limbo, not doing a proper job, so I was glad when the time came to start my engineering training at the Royal Naval Engineering College. I went home, with two other mids, in Landing Craft (Tank) 4039. It was a most delightful voyage at six knots from Malta to Portsmouth. At that speed and close to the sea surface there was always plenty of marine life to be seen. At one stage we fancied fresh fish so stopped a fishing boat and swapped coffee for fish at very favourable rates.

In the Bay of Biscay it blew up a bit and we had to resort for a while to the LCT survival trick of flooding the tank deck to ease the slamming of the flat bottomed vessel. Altogether a worthwhile passage home.

6

Off to College

The Royal Naval Engineering College was then at Keyham, near Devonport, and it backed on to the dockyard, which was convenient for practical training. However, we were to be its last inhabitants because a new college was nearing completion at Manadon, a few miles north of Plymouth. The new college was a hutted camp arranged around Manadon House, where the staff lived. It was reminiscent in a way of HMS Fisgard especially when trekking from hut to dining room in typical West Country wet weather. Not only was it a new college but we were to do a new, more academic, course that was based on the Cambridge University Mechanical Engineering Sciences tripos, with a few naval additions. We were also to do this in two years, not the usual three. This wasn't quite as bad as it sounds because we didn't have long vacations and we worked a six-day week and most evenings. Even so it was a bit of a slog and there were none of the diversions of university life, no Footlights or similar societies. It was sometimes cruelly said, usually by Cambridge people, that at the RNEC when the lecturer said, 'Good morning', the students wrote it down. Needless to say lectures were not optional.

All the same I did feel that if life was at times a bit grim and earnest it was purposeful, usually interesting and it was still possible to enjoy it. I worked hard, perhaps rather harder than necessary but I've no regrets about that. Despite the hard work my academic results were unexceptional; I was about average, in the middle of the class and remained so. I didn't shine in any other field of activity either. I was content to maintain a low profile and avoid drawing unnecessary attention because we 'officers under instruction' were under constant scrutiny and regularly made aware of our shortcomings. Mine included unruly hair, an incorrect salute and failure to attend church. There were almost certainly others but these I remember. Church was no longer compulsory so I rarely went. It was no great matter of principle and I did not feel that it should be a cause of adverse opinion. I was, I regret to say, no sort of rebel. Rebel or not I was soon to find myself very much at odds with authority and very surprised to be so.

It began on the occasion of our promotion to the rank of Sub Lieutenant when we needed an addition to our uniform outfits called mess undress. This was an expensive piece of kit and we were given a lump sum of, I think, £80 to buy it. In those days £80 was a lot of money and so very welcome. Whatever the sum was, when I came to collect mine I had what the Navy calls a North Easter. In other words, 'Not Entitled'. Naturally I asked why and at first no one seemed to know so I hoped it was merely an oversight. No such luck. The reason given was that I had already had a uniform allowance when I became a cadet and therefore had no entitlement to another. This was a severe blow and it really knocked me back with worry about how to manage financially. Very soon, though, I began to feel a little less glum because one of my less affluent colleagues confided that he too had had an allowance when he became a cadet but he had still received the £80. In addition I now recalled that I had a letter from the Admiralty when I moved from Apprentice to Cadet saying that my previous conditions of service no longer applied and that my new conditions would be in all respects those of a direct entry officer. Obviously someone had misread the rules and it could easily be corrected. Relieved, I sat down and composed a very proper official letter (Sir, I have the honour, etc., etc.) addressed to the Captain of the College pointing out the anomaly and requesting that it be put right.

Not unexpectedly I heard nothing for a week or so and then had an unwelcome reply. The rules had been correctly interpreted, it said but if I was not happy with this decision the matter could be referred upward to the C in C. The whole thing was so clearly a cock-up that my response needed no reflection. Yes, please, I said.

When the C in C's decision came it was all that I could have hoped for: There was indeed a misinterpretation of the regulations and furthermore one difficult to understand as my entitlement was plain. It concluded by saying, in effect, 'give the lad the money now'. This was done and happiness restored. There was, though, a nasty parting shot. The Staff Officer in charge of our course sent for me. He was a pompous sort of chap, his pomposity now enhanced by obviously strongly felt displeasure. I should not, he said, feel that I had won some sort of victory. What I had done was disgraceful; I had complained like some bloody-minded lower-deck lawyer and it was altogether un-officerlike.

It took me considerably aback and reminded me of the Captain of Dartmouth's more kindly expressed comments. On balance, though, the blow was outweighed by the lifting of the burden of financial anxiety.

Among the more exotic items of training equipment at the College were three Tiger Moth training aircraft to provide what was loosely described as 'air experience'. This was entirely voluntary, which had the beneficial effect that we few who were interested had them entirely to ourselves. Our instructors were those Staff Officers who happened to be pilots and we flew from the nearby Roborough airfield. I loved the Tiger Moth flying and became reasonably good at it though we could not fly solo because our instructors were not qualified as instructors and our lessons were unofficial. I did eventually go solo but it was nearly 60 years later in a Cessna 152.

I am biased, of course, but I believe that the RNEC provided an excellent, broadly based basic engineering training comparable with, or better than, any other British establishment at that time. It's gone now. Naval engineers go to university, take three years at it and then have to do additional courses to bring them up to standard.

Be that as it may, I sat my final exams with head humming with thermodynamics, fluid dynamics, physics, chemistry, metallurgy and even a little mathematics – and passed. Not outstandingly but as usual somewhere about the middle. It would do. I found, in fact, that I never used much of the laboriously acquired knowledge and over the years it leaked away so that now there is very little left. In exchange I garnered a rich store of engineering low cunning.

7

Driven by Steam…

We now began the last phase of our basic engineering training. Whatever our ultimate specialisations, we all had to qualify as Engineer Officer of the Watch (EOOW) in a steam warship. I was sent to HMS Indomitable to get my ticket. Indomitable was a fleet carrier of the Illustrious class. She had three engine rooms and three boiler rooms with six boilers, which could provide 110,000 shaft horse power on a good day. Indomitable had had a busy and eventful war. Quite early on she ran aground near Jamaica and was badly damaged. Then, while escorting a Malta convoy, she was hit by three heavy bombs. Later she took a torpedo in the port boiler room. Transferred to the Pacific she then received two Japanese Kamikazes. By the time I joined her all this excitement had taken its toll and this showed in some subtle and some very obvious ways. For instance, if on a calm day in harbour one stood on the flight deck aft and looked forward, the ship had quite clearly a list of two or three degrees to port. If one then walked forward and looked aft there was an equally distinct list to starboard. This twist had no discernable effect on performance. Down below we had the luxury of a machinery control room, an innovation for its time. Situated below the waterline and under armour, it had instrumentation showing what was happening to all the significant machinery in the ship so that the EOOW could see at a glance if any problem occurred and telephone instructions to deal with it. The flaw in this theory was that the telephone systems, like everything else, had been traumatised by the wartime experience and were liable to unpredictable failures. One therefore had, on occasion, a splendid overview of a developing emergency while lacking any means of dealing with it. One of the regular varieties stemmed from the ship's marginal generating capacity. Despite six turbo generators and four diesels, equipment added during and after the war came close to using all the capacity and if any of the generators was under repair or maintenance, things could and did quickly get critical. This resulted in generators successively tripping out under overload. As lights dimmed and the pervasive hum of fans died, the Commander was known to order all off-duty engineers to clear out of the wardroom.

It was, I suppose, good experience. There were admittedly very few dull moments and I duly got my Certificate of Competency which meant that I could take sole charge of the 110,000 horses and the telephones. I also got an additional job and became the Double Bottom Officer.

This unpretentious title covers rather a lot of responsibility in a ship such as Indomitable. The double bottoms comprise the space between the outer skin of the ship's hull and the inner skin, which lies across the bottom of the ship and also extends some way up the sides. The space between these two is about four or five feet deep across the bottom but may be wider in places up the sides. In this space the essential life-sustaining fluids for an aircraft carrier are stored, such as fuel oils, avgas (aviation gasoline), drinking water and boiler feed water. These spaces are not easy to access and even when empty are very difficult to crawl about in. You wriggle through the ship's frames via so-called limber holes. It's a bit like caving; you can easily become disoriented and if you have any tendency to claustrophobia this will find it. As might be expected these unpopular venues are rarely visited and then only by those whose duty requires them to do so.

The double bottoms are also the home of the valves controlling the underwater openings and the layman may be surprised to learn that in Indomitable there were over 300 of these. Due to all the wartime damage and repairs and a variety of hasty modifications, the ship's drawings had been poorly updated and in some areas completely lost so that some valves were not where the drawings said they were and others lurked secretly, their very existence unsuspected. I decided to visit all the known underwater valves in my realm as a check against the drawings and in the expectation of making rare and wonderful discoveries. However, I soon found that this was too ambitious an undertaking, partly because of the difficulty of access and the time it took but mainly because a more pressing concern became apparent – the poor state of the avgas system and the inadequacies of the system drawings. It had become routine to get round the problem by only using those parts of the system for which accurate drawings existed and which were known to be in good shape. However, this often imposed operating limitations that led to the reluctant use of the less trustworthy bits of the system and this sometimes showed up unsuspected defects in the form of petrol leaks. Petrol leaks were very serious and even small ones triggered emergency action. During the war a carrier, HMS Dasher, had blown up and sunk in seconds because of a petrol leak. So it was decided that the Avgas Officer and I should concentrate our efforts on physically tracing all the 'lost' areas of the petrol system and updating the drawings. We had this well in hand when the time came for me to move on but the job was not finished. Some months later Indomitable suffered a major petrol leak that resulted in a massive explosion and fire in which there were several fatalities. I don't know the detailed causes but it came as no great surprise to me.

Now the time had come to put the steam world behind me and get on with my specialisation as an aircraft engineer. There was no regret on my part. Much of it had been interesting and some had even been fun but the engine rooms of a warship do not in the least resemble that of the Woolwich ferry or indeed that of any merchant ship. Polished brass and majestic connecting rods in cathedral-like spaces are not the norm. The machinery is below the waterline, crammed into the minimum possible compartment and characterised by intense noise and almost unbearable heat. More important was the impression that steam was at the end of a line of development. It was getting old and set in its ways. The future was with aeroplanes; steam, for me was a thing of the past. How wrong I was became clear 11 years later.

8

In the Fleet Air Arm the Prospects are Grim

On the first day of my air engineering course we had an introductory lecture during which it was explained to us that we had to unlearn the crude methods to which we had become accustomed in the primitive world of steam machinery. We were now in the uplands of high technology and dealing with the delicate and temperamental products of the aviation industry. We were then taken from the classroom to make the acquaintance of some of these products. As we came round the corner of the hangar we were confronted by a Sea Fury and seated astride it an artificer belting away with a large hammer at something inside the engine. I found the sight curiously comforting.

I soon found that aerodynamics was not the most amenable subject for someone with no natural ability in mathematics, as it is a mathematician's paradise. I got by as I had in the past by learning what I could by rote and applying standard equations, accepting that I would never understand the beauty and subtlety of their derivation. Later – much later – I was to discover that mathematical ability and dexterity do not necessarily grant any insight into the beauty and subtlety of airflow nor encourage the intrusion of original thought.

Whatever my shortcomings in aerodynamics I finished the course and passed the final examination with, as usual, an undistinguished grade. It was sufficient. I was turned loose upon the world of naval aviation. What follows is a brief reminiscence of some of the less mundane events and memories of a small and peculiar world that has already slipped away.

I had begun my aircraft engineering career at a transition point for the Royal Navy and the Fleet Air Arm in particular. The old Navy, the executive branch at least, had not completely accepted the lessons learned about the supremacy of carrier-borne air power so convincingly demonstrated by the Americans in the war in the

Pacific. There was still a school that yearned for the pre-war battle-ship days and HMS Vanguard was still the senior ship of the Royal Navy. The Fleet Air Arm was, by now an accepted integral part of the Navy but in too many quarters the acceptance was grudging. The Fleet Air Arm itself was going through a confusing process and coping with uncertainties both operational and technical. The end of lease-lend meant that the magnificent American naval air-craft supplied on loan during the war had to be handed back or, reluctantly and sadly, scrapped. We were left with the legacies of the years before 1938 when the RAF supplied naval aircraft. These were the spawn of specifications written by men with no knowl-edge of naval aviation and produced by manufacturers of the same mindset. Such aircraft as the Barracuda and Firebrand were still in service and flown by brave men long after they should have gone to the scrapheap

Jet aircraft had been in service for six years or more but the Fleet Air Arm had none and there was some conjecture that such air-craft would never be operable from ships. Popular legend had it that this opinion originated from a report written by a senior naval test pilot in which he expressed his doubts that the average Sub Lieutenant would be able to cope with the demands of jet-powered deck landings. True or not, the average Sub Lieutenant never saw the report and unaware of his shortcomings, got on with the job. But that was still to come and meanwhile the navy operated obso-lescent piston-engined types such as the later Seafires, the Sea Fury and the dear old Firefly. The accident rate was high and casualties also, by today's standards. The latter two aircraft were even taken to war, in Korea. However, the tide was turning for the Fleet Air Arm. Despite the legacy of neglect and political indifference, despite poor equipment and inadequate funding, it was a highly motivated and self-confident organisation. Over the next 10 years or so it evolved into the force that originated all the innovations that made modern carrier flying possible, which provided the air cover for nearly all the operations undertaken by British forces and which earned the respect of its peers around the world while forgotten at home.

It was a very interesting time.

I have mentioned before that as a matter of personal policy I generally found it best to keep any inventions or ideas to myself.

However, very occasionally, if the idea seemed self-evident and simple I broke the rule. In most cases this proved to be a mistake. For instance, when I was the Air Ordnance Engineer in HMS Eagle I invented an improvement to the fusing units used on many of the aircraft of the day. There was a design deficiency that could cause bombs to be dropped SAFE when you wanted them LIVE and, much worse, would drop them LIVE when you wanted to jettison them SAFE in emergency. My invention was a simple modification involving a few minutes' work with a drill and a file. We tried it out. It worked. Every time. So we modified all the fusing units on board and had no more trouble. Except that as procedure dictated we advised higher authority of what we had done and politely suggested that they do the same. After a merciful delay we had a reply stating that the modification was quite unnecessary because Farnborough was developing a new fusing unit that would be electronic and much better. Accordingly, it went on, our fusing units were to be returned to their original condition. This gave us a problem because we couldn't un-modify them.

So we had to demand another set of units to replace them and this upset a tidy mind of some power who expressed the displeasure of authority at my waste of public money. Meanwhile many a bomb behaved other than as intended while we waited years for Farnborough's improvement.

Life on the flight deck can clearly be very stressful but it has its lighter side and, occasionally, moments of pure joy. One such moment came about while I was the Flight Deck Engineer Officer of HMS Victorious. I was responsible for the health of the catapults, arresting gear, lifts and a variety of other flight deck machinery. One day we were replenishing at sea and so there was no flying, which gave me the opportunity to catch up on some routine maintenance. Accordingly the deck plates were up just aft of the starboard catapult and a small team of engineering mechanics (what used to be called stokers) were greasing and checking the machinery that centred the aircraft as they taxied up to the catapult. I saw the Admiral descend from his eyrie in the island and approach the hole in the deck so I kept within earshot, being uneasy about admirals on my patch. The leading hand in charge came smartly to attention and bestowed an equally smart salute as the Admiral peered into the hole. He asked what they were up to and then, how did the machinery work. The leading hand gave an exemplary, succinct account of its function and there was a pause. 'It's too complicated for me,' said the Admiral. 'I can't understand it.' The leading hand smiled kindly, 'That's quite all right, sir – you don't have to.'

We all make mistakes and an engineer's mistakes can be expensive or worse. For that reason we engineers tend to be cautious and averse to precipitate conclusions. Nonetheless you can still achieve embarrassing errors despite deep thought backed by experience. I was once the Engineer Officer of a squadron operating Sea Venoms (among others). One day a Sea Venom approached the catapult and I noticed that a small panel over the wing fold joint had become crumpled and was sticking up unsecured. I stopped the launch and put the aeroplane unserviceable until the panel was replaced. These panels quite frequently did this due to poor design and my decision was considered to lack the 'press-on spirit'. Other squadrons, it was said, kept bolt cutters by the catapults and, when the panel misbehaved, cut it off and let the aeroplane go.

I was not happy about this because of the position of the panel and the discontinuity of airflow caused by its absence, which I suspected might result in a premature stall. I was very aware of two recent occasions when Sea Venoms had mysteriously been lost at night having apparently flown into the sea immediately after launch. The Senior Pilot thought that my misgivings were justified and we agreed to put the matter to the test at the next opportunity. So when we next disembarked we removed one of the offending panels and the Senior Pilot took off at a very substantial margin over the usual lift-off speed. At a good, safe height he made a gentle approach to the stall bearing in mind that Sea Venoms were rather variable in stall speeds and characteristics. This one proved to be more so than usual. Before he expected it and with no warning a wing dropped so suddenly that his head hit the canopy and the aeroplane was almost inverted before Jock caught it. We considered this completely convincing and never used the bolt cutters.

Fast forward now to another flight deck and another aeroplane, the Buccaneer – an excellent aeroplane in many ways but somehow it had inherited a wing fold panel with the same unfortunate habits as that of the Sea Venom. The first time the panel misbehaved I persuaded the squadron Air Engineer Officer to put the aeroplane unserviceable. Again I was accused of over-caution but it was reluctantly agreed that this would be the procedure until an air test could be arranged. In due course the test was done and I became instantly very unpopular because the absence of the panel made not the slightest difference to the Buccaneer's stalling characteristics. This is the sort of thing that makes your name in undesirable ways.

I suppose many people would imagine that engineering is a profession in which logic is supreme and deduction is standard practice and, very largely, this is so. However, there are times when due weight must be given to a funny feeling and notice taken of tiny things.

One fine afternoon while I was the Air Engineer Officer of 831 Squadron I became fed up with the paperwork and decided to take a walk through the hangars and out to the line to get some fresh air and remind people that I was about. On the way I met Lt B— who had landed from a sortie about 45 minutes earlier. He told me that during his last trip his right foot had got uncomfortably warm. 'I haven't put it unserviceable,' he said, 'but I've told the Chief Aircraft Artificer.' Lt B— had a reputation for this kind of thing, finding a possible fault that he wasn't prepared to put on record and which no one could reproduce. Even by his standards, though, this one was a bit unusual and mildly amusing. I ambled into the line office where the Chief AA reigned among all the aircraft maintenance documents. 'I gather Lt B— has had a hot foot.' 'Yes,' said the Chief with a long-suffering air, 'I've had a look at the aeroplane and given it a run. It's OK and it's on the line for night flying.'

Quite suddenly I had a funny feeling about Lt B— and his hot foot. The funny feeling wouldn't go away until I had had a look at the aeroplane myself. The situation was delicate; the CAA had done all that anyone reasonably could. He had investigated and run up the engines and found no fault. He was competent, reliable and a man I greatly respected. He was almost certainly right. Nevertheless I wanted to look at the aeroplane to make the funny feeling go away. 'I'll just stroll over and take a look at it.' I said. His face became deadpan; he had every right to be offended. To soften the blow I smiled and said, 'We must demonstrate that we take these things seriously.'

In the Gannet aircraft the pilot sat above the engines so I sent for a ladder and unclipped the engine cowlings on the starboard side. I had a good look round but I could see nothing amiss at all. I gave it all another look but the situation was unchanged. There was nothing wrong. I paused for a moment, thinking about how I would talk about it to the Chief. I closed up the cowling and got down feeling annoyed with myself, I had needlessly upset the CAA for nothing more than a funny feeling. The trouble was that the funny feeling was still there. I went back up the ladder and unfastened the cowling again. I had the inside of the cowling in front of me and there was a pencil line drawn down it. I wondered, idly why someone would draw such a line down the inside of the cowling and rubbed it with my finger. It smudged, it wasn't pencil it was a very precise, sharp-edged deposit of very fine dust. A light was beginning to dawn. I examined the engine at the positions corresponding to the dust mark on the cowling but there was nothing to be seen.

However, when I ran my fingernail across the position there was a great deal to be felt. The compressor casing was cracked over about half its circumference.

That engine could have failed at any time and although the Gannet had two engines if the failure had happened on take-off or during landing it could have been fatal. We were lucky. The CAA was particularly upset, though he had nothing to reproach himself with, and I gained an undeserved reputation for extra sharp eyes.

As well as funny feelings it's unwise to ignore funny noises. There is always a reason for them, usually a trivial one, but sometimes they originate from a distinctly unamusing source.

It was quiet in the hangar – a very rare condition and due to the fact that we had finished work on the aircraft. They were all serviceable and ready to fly except for one Gannet that was awaiting spares. Three Gannets and four Sea Venoms would be out on the line when the CO appeared, which wouldn't leave him much to complain about.

A naval air mechanic sat in the cockpit of one of the Sea Venoms waiting for the tractor that would tow the aeroplane out to its place on the line. His job was to operate the aeroplane's brakes. While he waited he idly moved the control column about in a semi-comatose manner. He was tired, we all were. All at once he became more alert, he tilted his head and stirred the control column more vigorously, he kicked the rudder bar fully each way and then sat there, deep in thought. After a few moments he got out of the cockpit and walked over to the line office. 'Chief,' he said, 'there's a funny noise' – and everything was changed.

I was in the line office going through the forms A700, which record in detail all the maintenance and repair work done on the aircraft, with the Chief Aircraft Artificer. I took no particular notice of the air mechanic's report as the CAA excused himself and went to look into it. He was away quite a while and returned looking very puzzled.

'There's a funny noise when you move the pole or the rudder bar.'

'What sort of a noise?'

'It sounds as if someone is using a ratchet screwdriver in there.'

I went over to the aeroplane, got into it and waggled the pole. The sound came from behind the cockpit and sounded, as the Chief said, like a ratchet screwdriver being used. It wasn't very loud, in fact you probably wouldn't notice it at normal working noise levels unless you were listening for it, but now that we knew it was there it could also just be felt as a slight roughness when the control column was moved. The thing was, what was it and did it matter, because investigation was not at all straightforward. The sound came from behind the bulkhead at the rear of the cockpit where all the control cables ran across the fuselage in a narrow gap bounded to its rear by the front bulkhead of the engine bay. Our problem lay in the fact that there was no access of any kind to this space, no inspection panel, no spy hole; it was a completely inaccessible part of the fuselage structure closed from the day it was built.

Did it matter? We checked the other three Sea Venoms. One had a very faint rubbing noise but the others were silent, otherwise no helpful clues. We checked the control friction loads. The noisy one was on the high side but well within limits; cable tension was spot on. No clues there either. At this stage no one was venturing an opinion although it was quite clear that whatever it was the control cables were involved. There was only one way to clear up the mystery and that was to break into the space. That, in itself, was not too difficult. The Sea Venom cockpit was built of wood, like the wartime Mosquito, and the rear bulkhead was plywood. It was simply a matter of taking the ejection seats out and making a suitable hole. But this was a serviceable aeroplane; there was nothing wrong with it except a funny noise – and not a loud one at that. Suppose we broke in and found nothing? The CO would not be pleased at being deprived, unnecessarily, of a perfectly good aeroplane, though that was the least of my worries. My reputation would be in tatters and the tale would lose nothing in the telling. 'He hears this funny noise so he grabs a hammer and chisel and knocks a hole in the bulkhead.' Mirth all round.

It was up to me and I did not want to make any kind of decision particularly at 3 a.m. and having been awake (more or less) for 21 hours but there was no avoiding it. 'Open it up, I'm going to bed, call me when you can see inside.'

The call came at dawn and when I got to the Squadron the Chief's face was all the assurance I needed. 'You'll never believe it,' he said. Certainly, even with the evidence of one's eyes what we had found was astonishing not least because we were unable to figure out how things had got into the state we found. Because instead of neat parallel rows of control wires running across the fuselage there was a great Gordian knot of wires twisted around each other yet somehow able to move and to actuate the correct control surface in the correct sense. The rubbing of the wires together made the ratcheting noise and had caused some wear on the cables, though luckily not yet enough to fray them. The big question was how had things got into this state? A thorough check of the aircraft's documents showed that the control cables had never been disturbed in service and that there could be only one incredible conclusion – it had left the factory in that state.

Now was this an isolated incident or were there more like it? Emboldened by now, I ordered the Sea Venom with only a faint rubbing noise to be hacked open, which revealed an even more spectacular knot, albeit a quieter one. This was because the cables had worn smoother though still not frayed. This aircraft had also left the factory in this state.

I flashed off an Important Aircraft Defect signal grounding all Sea Venoms while we checked our last two Sea Venoms to find that they at least were as they should be but of course our four serviceable Sea Venoms were now zero. The CO was not pleased but was advised to direct his displeasure towards the De Havilland Aircraft Company. Over the next few days it became clear that about one in three of all Sea Venoms were in this state but we never did find out how they got that way. One wonders how long the situation could have gone on and whether it would ultimately have been discovered the hard way. Because the lad who heard and reported the noise said afterwards that he really thought that it wasn't anything to fuss about and he very nearly ignored it. Funny noises are treacherous.

Of all noises nothing grabs the attention so firmly as a sudden loud bang when airborne. The not unreasonable assumption is that the SLB is the precursor of imminent disaster or at least of something dangerous and unpleasant. I can report, with gratitude and humility, that it is not necessarily so. My first SLB experience was at Boscombe Down when I was a Trials Officer there. The aeroplane was a Fairey Gannet and we were carrying two 500lb bombs (inert, full of sand) one under each wing, which we were to release at Vne over the range at Lyme Bay. (Vne is the maximum permissible speed of an aeroplane, often only achievable in a steep dive, in this case, something like 350 knots.) So there we were, howling steeply downwards waiting for the speed to reach 350 knots when BANG! and a thump felt through the airframe. WHASSAT? WHASSAT? and an immediate throttling back while pulling out of the dive ever so gently. As gallons of adrenalin began to subside we looked around for a cause but could see nothing and instruments were normal but we felt it prudent to return home. After all something had probably cracked and was just waiting to fall off – a wing perhaps. We landed and nothing fell off so we got out and had a good look round. When we found the cause it was scarcely believable. The bombs are carried on special carriers that hold them tight under the wing. These carriers are usually of a messy, unstreamlined shape so they are covered by light fairings to reduce the drag. In this case the nose fairing was, in essence, a light alloy hemisphere about 18 inches in diameter and under the air load it had collapsed like so much screwed-up paper. Who would have thought it could make such a bang and thump.

The next SLB was during a maintenance test flight in a Grumman Avenger. I had two of these relics in 831 Squadron, they were lease-lend leftovers and I loved their honest, sturdy Grumman airframes and their beautiful Wright R-2600 radial engines. We got airborne for a routine test flight following some heavy maintenance and were still climbing to a respectable height when BANG! This time the gallons of adrenalin were still circulating when BANG! AGAIN! These bangs occurred every few minutes accompanied by heavy thumps. Nothing fell off so we were able to take a look around. Being an old-fashioned aeroplane the pilot had a proper pressure gauge for the hydraulic system and it became obvious that each bang was attended by wild fluctuations in hydraulic pressure. So home we went and despite the hydraulic bangs and thumps the wheels and flaps went down and we landed with no trouble. It turned out that there was an internal leak in the hydraulic system that allowed fluid to seep back to the tank. When, eventually, the hydraulic accumulator was emptied it immediately put the hydraulic pump on full stroke to recover the pressure. Who would have thought that such a thing could cause such a bang and thump? Despite my care and attention I didn't have that lovely Avenger for much longer. One day we sent it to Fairwood Common to pick up some spares. It was in the hands of Lt B—, he of the hot foot mentioned elsewhere, and Lt B—was suffering from a monumental hangover. Whether that is relevant or not the fact is that he landed the aeroplane with its wheels up and claimed in mitigation that he was distracted by the air traffic controller shouting, 'Wheels, wheels!' at him. I never forgave him.

Naval aeroplanes lead a hard life and tend to show their age, some more so than others. Sea Venoms did not age gracefully. The forward part of the fuselage was a plywood and balsa sandwich about 20 to 25 mm thick. This was covered by a layer of heavy fabric doped onto the wood. To be fair the wooden bit was no trouble, in my experience, but the fabric covering soon crazed and cracked and looked very unsightly. Cracking was not confined to the superficial covering either; some metal parts in the wing fold area were prone to it and had to be checked frequently. As a result any middle-aged Sea Venom with its apparently severely cracked fuselage and the grubby residues of pink crack detection fluid adorning the wings was not an attractive sight. Scruffy was the only appropriate word.

One autumn I found myself and my scruffy Sea Venoms at Andoya in the Lofoten Islands operating from a remote Norwegian air base. I'm not sure what we were doing there but, on return from their sorties, some of the pilots liked to do a low-level high speed run over the airfield. I think this was discouraged at home. Anyway, on this occasion we watched the approach of a Sea Venom at few feet and many knots when, to our momentary dismay, it appeared to disintegrate. Pieces seemed to explode from it but thankfully the aeroplane flew on, seemingly in one airworthy piece while the other bits wafted gently to earth where they were revealed to be chunks of fabric. The aeroplane, on landing, with its wooden walls exposed, looked as if it might have been acceptable in Nelson's navy.

It certainly wasn't acceptable in ours so we sought means of covering its nakedness. The Norwegians had little to offer as this was not a main base and had few facilities but they did their best for us. We found some lightweight Madapolam fabric intended for gliders and some clear dope. What these were doing so far north of the Arctic Circle we didn't ask. I decided, on no firm grounds, that three layers of the glider fabric would be the equivalent of the original when applied with four layers of the clear dope and that was done with only hours to spare before we flew home. I liked the look of the aeroplane with its clear doped finish, which was like antique, lovingly polished, varnish but that had to be hidden fast, as soon as we got home, under a coat of the standard naval aircraft finish.

After a week or two we began to realise that the aeroplane's finish was not cracking and crazing. It remained beautifully smooth and shiny, even better in fact than a new aeroplane and my CO began audibly to wonder why the others couldn't be the same. Ever anxious to please higher authority I gave all our Sea Venoms the same treatment over the next few weeks as opportunity occurred.

We had the smartest, cleanest Sea Venoms in the Navy and people noticed. Some swallowed their pride and asked how it was done, offering liquid reward for the information but we affected an infuriating smugness and claimed that it was the product of good husbandry and sound engineering practice.

The relationship between a squadron Engineer Officer and the aircrew, especially the CO, is a very delicate one. On the one hand the aircrew are risking their necks on a daily basis while the engineer is in relative safety on the ground or flight deck. This discrepancy can be mitigated to some extent by the engineer volunteering to fly on maintenance test flights, for instance, but this is obviously not possible with single seat aircraft. Also, with modern military aircraft, getting dressed for the occasion is a protracted affair while everything is fitted and adjusted. This wastes a lot of time if you don't fly regularly. And if, like me, you have difficulty clearing your ears during fast descents from high altitudes you are letting yourself in for severe pain and weeks of slowly decreasing deafness.

Most aircrew display a touching faith – 'If you say it's serviceable, I'll fly it' – but it comes with unspoken caveats. If you say it's serviceable but you have made an error you will not be easily excused. And if that error is thought to be negligent there will be no mercy; the discrepancy of safety comes to the fore. Aircrew make mistakes too but feel able to tolerate them because the neck at risk is their own. Engineers' errors put someone else's neck in danger. Understandable, therefore, that there is an underlying uneasiness and, if the engineer is wise, a slight distance except for a close alliance with the senior pilot.

The CO and the Engineer Officer do not necessarily have mutually compatible objectives. A squadron CO is an officer on the brink of promotion – provided that he makes no serious errors and that the squadron performance under his leadership is at least as good as under his predecessor. Ideally it should be better, much better to ensure promotion. The temptation to force the pace is unlikely to be resisted especially while the 'press on' spirit reigns supreme. It is in this context that the CO and his Engineer Officer are likely to find themselves at loggerheads because few people understand the random nature of defects in complex equipment and a run of bad luck in aircraft serviceability may generate a dark suspicion that the engineer is not pulling his weight or being technically over-exacting. The difficulty in proving or disproving such suspicions can give rise to an atmosphere of mistrust.

The engineer, for his part, needs a good report from his CO and this involves demonstration that he is a dedicated adherent of the 'press on' spirit while avoiding any serious technical failures. This can mean some heart-searching engineering judgements because it is easy, and safe, to ground an aeroplane but not admirable. I walked this tightrope and must have got some things right because the squadron received the Boyd Trophy and the CO was promoted – but there were some emotional episodes along the way.

One such episode was during a NATO exercise. All four Gannets had flown during the forenoon though one was now on the ground for a mandatory inspection that would keep it there for at least eight hours. The CO was not at all happy about that and made it clear that he wanted all the three remaining for the night exercise to follow. That must have seemed a reasonable expectation. They had landed serviceable from the previous sortie and were not flying again until that night. Night duly arrived and the aircrew strolled out to man their aircraft, the pilots doing their walk round inspection before getting into their cockpits. One of them didn't bother getting in; he poked his head in my office to say, 'I'm just going to put my cab U/S – looks like we've got another gearbox failure.'

I went out to look at the aeroplane although I knew what I would find. Gannets had two co-axial propellers each driven by a separate engine via a common gearbox. The port engine drove the front propeller, a blade of which I seized and swung. It moved with considerable resistance and a grinding sensation. Turboprop propellers should move smoothly and with light resistance. This aeroplane wasn't flying tonight. I gave instructions for it to be towed in and start changing the engine. I had only been back in my office a few minutes when the CO appeared to tell me that he had countermanded my instructions and wanted an explanation for them. I thought demonstration would be the most compelling argument and asked him to come with me to the aeroplane where I went through the business of swinging the propeller and invited the boss to do the same.

He did so and said that some roughness and resistance was to be expected, that the pilot and I were being over-cautious, that this was an important exercise and we were a front line squadron not a training outfit. He wanted to know exactly what was wrong with the engine. I said that I reckoned that there was a bearing failure in the gearbox. He said that I reckoned, I didn't know. It was all supposition and he wanted the engine started right now. I knew that the engine wouldn't start and that the attempt to start it would probably wreck it completely so I said that first I would need to drop the oil filters and that I'd bet that the oil would come out looking like silver paint mixed with bits of bearing. This was done and the oil glugged out looking exactly as predicted. The Boss's frustration and bafflement were distressing to watch; fate and an engineer were conspiring to thwart his reasonable and noble objectives. Still, given time he mellowed. I didn't expect an apology but, much later, he introduced me to someone as 'my tame plumber who works miracles'. Perhaps he had in mind the transformation of clean oil into a mixture of silver paint and bits of bearing.

Some members of the steely-eyed warrior class are inclined to be suspicious of engineers, believing them to be deviously pursuing a less motivated agenda than the said steely-eyed and causing unnecessary obstructions with their petty technical problems. As a consequence they believe it to be essential to squash with the firm hand of authority any sign of deviation or dissent from a proposed line of action. My Commanding Officer at the RN Air Station, Yeovilton, was one such, when I was the Senior Engineer.

Every year the station had an open day when the place was thrown open to visitors and an air display was laid on. In addition to the air display there were a number of sideshows, as in a fairground, to encourage the visitors to contribute to naval charities. These sideshows were improvised from various bits and pieces by the staff of the station workshops and the design, manufacture, setting up and manning of them was the responsibility of the Senior Engineer.

A few weeks before the open day the Captain sent for me and told me that he wanted the centrepiece of the sideshows to reflect the fact that Yeovilton was now the home of the first squadron to be equipped with the new McDonnell-Douglas Phantom. Accordingly this sideshow should give the customer the impression of landing a Phantom on an aircraft carrier. This was a far cry from the usual swings and roundabouts and slides and I wondered if the man appreciated what he was asking for. I didn't wonder for long. 'Take a look at the Phantom simulator,' he said. I was appalled. I had seen the Phantom simulator, which was the size of several houses, crammed with miles of cable and tons of clever electronics and very realistic. It even went 'kerlunk, kerlunk' as the simulated wheels ran over the simulated joins in the simulated taxiway. This was a situation needing a tactful approach. 'What did you have in mind, sir?' I asked.

'I've told you what I want,' he said, 'so stop dodging and get on with it.'

The situation was beyond fear of consequences. The task was beyond me or anyone else for that matter and I hadn't the vaguest notion of what, if anything to do about it. I went to my friend the Deputy Electrical Engineer to see if he had any ideas but he knew a poisoned chalice when he saw it. He oozed sympathy as one does for a victim of plague but made it clear that he was not going to risk infection. I was on my own.

I decided to do nothing unless it could be done, like all the other sideshows, using spare bits and pieces of wood and metal. That was a useful mind-clearing decision because quite soon I had a brainwave and could set my team to work. We made an approximate scale model of a carrier's flight deck about 20 feet long and including five arresting wires. A raised platform a few yards astern of the flight deck supported a mock-up cockpit made from various bits of wreckage including a real (but not Phantom) control column (Figure 1). A sloping wire ran from the stick to a stanchion forward of the flight deck. Movement of the stick backwards or forwards increased or decreased the tension of the wire and so varied its height over the deck. Now if you attach a weighted model Phantom to the wire and allow it to slide down you can control the point at which it will 'land' on the deck. Prizes, a bag of sweets, to be awarded to those few who managed to engage number 3, the target wire. My lads enjoyed playing with it but we wondered how the children would respond on open day. We need

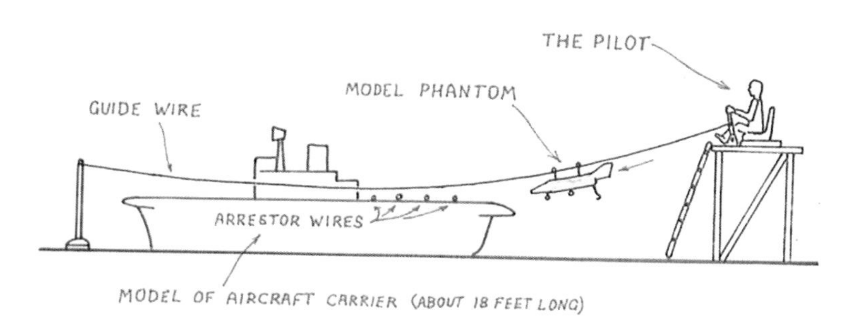

Figure 1 The Phantom simulator

not have worried; it was runaway success. Long queues formed of small boys and their dads and even a few of their sisters with takings breaking all records. Even so the Captain was not happy. 'That's not what I wanted,' he snorted. There was an interesting sequel a few years later when we had both left the Navy. There appeared in toy shops a smaller, table-top version of my 'Phantom Simulator' and I discovered that my ex-Captain was now a senior figure in the firm that made it. So far as I know it never caught on in this miniature form.

All of our country's defence equipment is procured through the medium of a large government department that changes its name frequently (and who can blame it). When I was sent there it was called the Ministry of Supply and I had the grandiose title of Deputy Development Project Officer (Buccaneer). My job was to ensure that the Buccaneer was supplied to the Navy fully in accord with the Navy's exacting standards. In order to do this I was armed with no authority whatsoever. Masses of paper, much of which I understood imperfectly at best, passed through my hands daily.

These were early days in the Buccaneer's life and aeroplanes and lives had been lost in test flying accidents at sea. It was considered that the crew should be able to use their ejection seats to escape underwater. This had been done successfully, in desperation, years before from another aircraft but it was not generally looked upon as a recommended method of escape and no one else had tried it. The Buccaneer presented special problems. One was that ejection was through the canopy, the ejection seats being provided with large prongs at the top for this purpose. The other was that trials showed that the higher power ejection guns used in the Buccaneer produced a lethal blast wave underwater. The solution proposed by Martin-Bakers, who made the seats, was to use a separate lower powered ejection system underwater that used compressed air. The trouble was that this added weight and complication. Finding the space for the compressed air bottles was not easy and the canopy would have to be jettisoned because the air system could not break it.

At this stage I had a brainwave and devised a fairly simple modification to the ejection seat that would allow the normal ejection handles to be used to fire the normal cartridges and break the canopy. At that point a port would open and allow the excess gas pressure to be vented harmlessly and relatively gradually, thus avoiding the dreaded blast wave at the end of the ejection stroke. I modestly considered this to be far superior to the Martin-Baker proposal. So much so that I forgot my principles and took the idea to my boss – Captain Dyer-Smith. He liked it, too, and told me to get it properly written up and drawn while he arranged a meeting with the seat makers. In the event the meeting was to be with the legendary Sir James Martin himself, a man who was known to express his views in uncompromising terms. According to my boss, Sir James had no love for the Ministry of Supply and with good reason – 'But don't let his manner put you off,' he said, 'he's all right.'

So off I went to Denham and was duly shown into Sir James's office. We soon got down to business and I was invited to explain my idea. He showed a great deal of interest and asked some penetrating questions, which didn't bother me, except that he had an unusual way with words and a pronounced Irish accent that made me wonder if I was getting my message across. However, it soon became clear that he understood exactly what I had in mind and for a while I began to think he approved. Eventually he grinned and said, 'Well, I'll show you what I'm going to do about this.' He gathered up my drawings and slowly screwed them up then threw them at the wastepaper basket, which he missed.

I was shocked, hurt and furious. To give myself time to think and to calm down I bent down and recovered my drawings and smoothed them out on his desk. 'Why did you do that?' I asked. 'What's wrong with the idea?' He grinned again. 'Nothing,' he said, 'except that you will patent this, won't you?' I told him that I would not but that the Ministry almost certainly would. 'Exactly,' he said, 'and that's the problem because nothing goes on to a Martin-Baker seat that doesn't have a Martin-Baker patent – now come and have some lunch.'

The compressed air system duly went into the Buccaneers but as far as I know was never used. Despite the disappointment I couldn't help liking him and, years later I was proud to receive the Sir James Martin gold medal for something entirely different.

One of my duties in the Ministry of Supply was to certify that the Buccaneers and the bits and pieces that went with them were delivered to the standard required and that the prices charged were 'fair and reasonable'. From the moment I arrived I was uneasy about this aspect of my duties because nothing in my training or experience gave me any expertise on the subject of the costs of manufacturing aircraft and their parts, nor how these costs could be expressed as fair and reasonable prices. My predecessor, who displayed, I thought, an unseemly haste to depart for his next appointment, played it down. He obviously reckoned that I was being a fussy old woman about a trivial matter. 'You'll have much more urgent things to think about than that,' he said. 'Just sign.'

He was right. I had a lot to do and when it came to prices I just signed and never had a moment's trouble. It was quite touching, the evident faith that my signature inspired, because never was it queried. No one came along and cried to heaven that I was a fraud who should be exposed and so I grew confident; but always lurking in the back of my mind lay the thought that somewhere in the Ministry there existed someone with the expertise that I so conspicuously lacked. I was not to know that this was not the case. So I signed and signed. One thing was abundantly clear: aeroplanes and everything connected with them were astonishingly expensive. Even a humble piece of waterproof material to protect a pilot's cockpit canopy could rival a Persian carpet in price. Then one day along came a mind-boggling item – or rather a set of items. Most military aircraft have the ability to carry a variety of items under their wings, such as bombs, missiles and extra fuel tanks. The fuel tanks can be jettisoned in case of emergency and are usually loosely known as drop tanks. The Buccaneer's drop tanks were specially shaped to fit snugly under the wings but otherwise were unremarkable apart from the price, which set new records so far as I was concerned. I was about to sign in accordance with what had become a custom when I had a naughty thought. What would happen, I wondered, if I didn't sign. I asked around and got a variety of opinions but one, more knowledgeable than most reckoned – 'No problem, it goes to Technical Costs and they sort it out'. So I sent it off to Technical Costs with a light heart.

Not unexpectedly nothing happened for a few days but then, suddenly, there was uproar – unexpected and unwelcome as far as I was concerned. It appeared that the manufacturer, hearing that there was a query about his price, had suspended delivery. For hours I endured mountains of verbal ordure falling upon me from a great height. It was made clear to me that whatever insignificant opinions I had regarding the price they carried no weight at all and that by invoking Technical Costs I had thrown an enormous spanner in the works. My orders were from on high and admirable in their clarity, 'Get that form back and sign it – just sign.'

So I learned what perhaps I should have known, that the business of calculating a fair and reasonable price for aircraft parts is not a laborious process of adding all the various moneys spent in producing the bits, adding the overheads and then adding a reasonable, but of course modest, profit. It is, in fact much simpler and the same as for anything else – the price is what the customer will pay.

I have mentioned before that the Royal Navy's attitude to engineers is, to be kind about it, somewhat mixed. Most of us treat it as one of those minor irritations to be ignored while getting on with life. Sometimes, however, it becomes rather difficult to take lying down, especially when it gets personal. The Executive Officer (or second-in-command) of HMS Victorious was a Commander of the seaman branch of the Navy. He considered himself to be a particularly well-educated and cultured man and saw no reason to disguise the fact. His exquisite sensibilities were particularly offended by engineers, who, he believed, were vulgar barbarians simply as a consequence of their profession. He saw no reason to disguise this view either. We engineers lived with it by summoning up our vast reserves of good grace. However, there came a time when prejudice descended into embarrassing farce. From time to time, on a Sunday night in harbour, the wardroom would lay on a so-called soirée to which local VIPs would be invited. This was all part of our 'showing the flag' duties. The preparation and decoration of the quarterdeck for this function was undertaken by the officers of the various departments in turn. Inevitably a degree of competitiveness crept in and when it came to the engineers' turn we felt that we had a few advantages. For a start we had inherited all the original ship's drawings, beautifully done, in colour, on linen. They were now obsolete, but laundered they made colourful and uniquely abstract patterned tablecloths. In addition we had inherited, from the days of petrol-burning aircraft, hundreds of brass wheel spanners that were no longer needed in these days of kerosene burners. Wheel spanners are really quite elegant objects consisting of a shaft that terminates in a delicately curved hook arrangement. None of your crude motor mechanics implements. Our plan was to polish these up and use them as table decorations and even as a chandelier. These things we did on the day and the whole thing looked splendid. With a final fond glance we went below to change into our mess kit to receive our guests.

Shortly after this, it appears that the Captain went ashore to collect his guests, remarking in passing to the officer of the watch that 'the plumbers have put on a good show.'

I had just finished changing when a messenger appeared saying that the Commander was demanding my presence on the quarterdeck forthwith. I found him quivering with outrage and barely coherent. I gathered that this was disgraceful, a vulgar and ignorant display and that these awful 'engineering THINGS' were to be removed before any guests arrived. So we turned to, with no good grace, and got rid of it all with little time to spare.

Meanwhile, unknown to us all, the Captain was on his way back to the ship having regaled his guests with a description of the splendid sight awaiting them on the quarterdeck. Not unnaturally he was none too pleased either. We would have loved to know what passed between him and the Commander but in due course we were informed discreetly that our effort was appreciated. These things are sent to try us.

The island structure of an aircraft carrier provides vantage points from which the various flying operations can be watched. These positions are known as 'the goofers', the occupants are 'goofers' and the process is 'goofing'. In the nature of things, space in the goofers is very limited but the spectacle is often well worthwhile. Back in the fifties it was always exciting. Landings in particular were more exciting than any motor race and there was always the possibility that the goofers might suddenly become part of the show. One recalls, for instance, the Wyvern that embedded itself in Eagle's funnel. As the years have passed, however, the excitement (and the casualties) have greatly decreased due to the advent of jet aircraft (from which you can actually see ahead when in the landing attitude), the angled flight deck and the projector sight. Deck landing accidents are now mercifully quite rare and all the more surprising when they do happen.

When you have watched a few hundred deck landings you begin to develop an eye for the process such that if it begins to look as if the approach will not result in a happy outcome you can remove yourself from the scene in a dignified and timely fashion. In Victorious, during recoveries, I stationed myself on the starboard edge of the flight deck between numbers two and three wires because this spot was adjacent to a relatively unobstructed sponson into which I could leap if things went awry. I had occasion to use it only once, an occasion of two surprises.

The first surprise was the landing itself, during the last recovery of the night flying. I had been watching the Sea Vixen's approach, which appeared to me to be quite normal, and I had perhaps allowed my attention to wander in the direction of my bunk, which I had not seen for far too long. Whatever the reason it very suddenly became clear to me that the approach was no longer normal. He was low – very, very low. It was time to jump. Brain said, 'Jump' but it took a nanosecond or so for the message to reach the legs and in the interval the Sea Vixen hit the round down in a great shower of sparks and an ear-splitting double bang as the undercarriage collapsed and the empty drop tanks burst. I felt something hit my chest and bits of aeroplane drifted lazily by. At this point the message reached my legs and I departed into the welcoming sponson. After a decent pause and having convinced

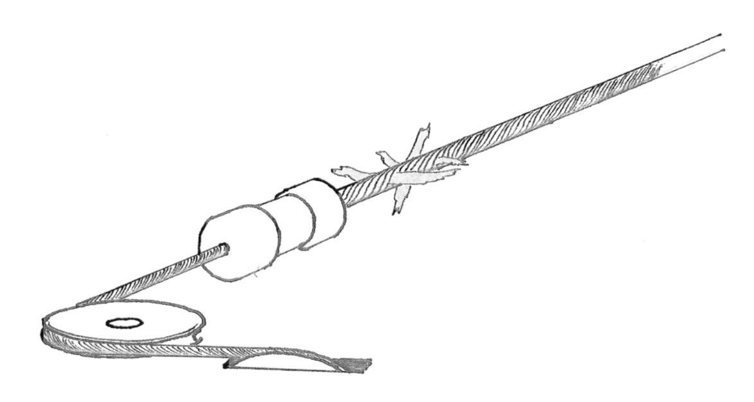

Figure 2 Arrestor wire 'pierced' by fibreglass

myself that I was alive and well I peered over the deck edge to assess the situation. There was a very bent looking Sea Vixen abreast the island but luckily no fire and the crew was emerging looking none the worse. So far, so good. Now to check the state of my arrestor wires and here came surprise number two. The aeroplane had caught one of the wires and the centre span (the bit that engages with the aeroplane's hook) was a most astonishing sight, because of its impossibility. The centre span was a steel wire rope about 35 to 40 mm in diameter yet at each end, near the dumbell connector to the main reeve, it was pierced by pieces of fibreglass drop tank, as shown in Figure 2.

Now anyone who has handled steel wire ropes of this size will appreciate that it is far from easy to drive a steel marlinspike through the wire with a hammer, so how could the wire be penetrated by relatively fragile fragments of fibreglass about 3 mm thick? The answer had to be that somehow the strands of wire had been separated sufficiently for the fragments to pass between them but the means by which this happened was still a mystery. We reported the mystery and were given the solution, which is as shown in Figure 3. It seems that when the arrestor hook engages the centre span a shock wave is created that travels outward until it hits the dumbell connectors. At this point it is reflected back into itself and this causes the strands to open up momentarily into a football-shaped and -sized basket before closing and trapping any passing bits of drop tank. A likely tale, you may think, but I have

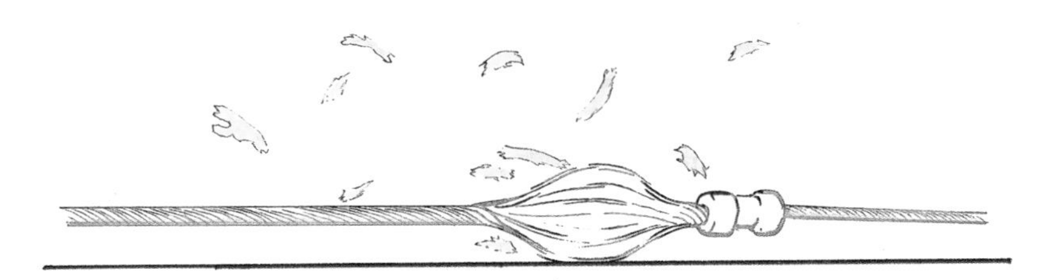

Figure 3 How it was done

seen the high-speed film and it is true. It must also be confessed that since the event I watched the wires during recoveries with an eagle eye but never did I catch them at it. This is the kind of thing that makes engineering so interesting, especially when you begin to think you've got it almost mastered. There's so much going on under your nose of which you are unaware.

9

The Aeroplanes

I first went to sea with Seafires and Fireflies. The Seafires were Mk47s, which were the last of the extensive Spitfire range. They were nearly twice the weight of the early Spitfires and a lot faster but they shared the characteristics of flimsy undercarriage and no view forward in the landing attitude. This made for some entertaining goofing but was expensive for the taxpayer. I only ever worked on a Seafire during training, enough to be sure that accessibility and ease of maintenance were not foremost among the designer's thoughts. The Firefly was very much the workhorse of the Fleet Air Arm in those days and sadly underrated. My first ever flight was in the back of a Mk5 and Fireflies went on for years, ultimately staggering into the air fat and grossly overweight as the Mk7. Mine is, of course, an engineer's viewpoint but pilots described the Mk5 at least as 'a gentleman's aeroplane'.

The Navy had some really bizarre aeroplanes in those days – a legacy of RAF control – and one of the most ugly of these was the Fairey Barracuda. Pilots' opinions of it varied from the frankly scared to 'it's not nearly as bad as they say'. In Indomitable we had two Barracudas as Ship's Flight, that is as general hacks for ship to shore and back, fetching the mail, etc., and one of the pilots was a friend of mine so I flew with him occasionally – always in the gunner's seat, never the observer's position under the high wing, because the Barracuda's immense undercarriage was prone to collapse. If this happened there was only a thin sheet of 24-gauge aluminium between you and the deck or runway. One Sunday morning when the ship was alongside in Gibraltar and the Barracudas ashore at North Front my friend and I decided to go on a sightseeing flight. You could do things like that in those days. We took off and went up the coast to Malaga then crossed the straits to take a look at Ceuta and Tangier. On the way back my driver announced that he was going to do a slow roll. I was convinced that the Barracuda was incapable of doing any kind of roll and that he was just trying to worry me so I laughed and took no notice. Then we rolled, slowly and barely barrelled, so that as we became inverted I started to slide from under the loose lap strap that passed for a harness. Naturally I put my hands up to the canopy over my head and grabbed the bar running across it whereupon the canopy swung open as it was designed to do. I was in no danger of falling out – nothing could have broken my iron grip of terror on that bar – but the relief on returning to level flight was almost worth the fright.

Rather less bizarre but with less excuse, because it couldn't be blamed on the RAF, was the Blackburn Firebrand. This was designed as a torpedo carrier with the pious hope that after dropping its torpedo it could become a fighter of sorts. Considering that it was designed from the outset as a naval aircraft there was no excuse for its having the longest nose of any, with the pilot closer to the rudder than the engine. To compensate in some degree for the complete inability to see ahead in the landing attitude there was an additional air speed indicator set into the outside of the port side of the fuselage for the benefit of the pilot, who had to crane his head out of the cockpit in order to see it. We had a squadron of Firebrands in Indomitable that provided the most spectacular deck landing manoeuvres I have ever seen. Other peculiarities of the Firebrand were the fin being offset by a considerable angle and the pronounced upfloat of the ailerons in flight. A much more respectable contemporary was the Sea Fury, the last and the best of the Navy's piston-engined fighters. It combined high performance with good range and endurance and a deck-landing undercarriage. Design for maintenance wasn't bad on the whole, either. While we are on that subject, the best aeroplane I ever came across for maintainability was the American F86 Sabre, which was proof that performance and maintainability can be combined if the will is there.

Then, rather late for the Navy, came the jets. From an engineering point of view things suddenly got much simpler but it was not to last. The first of these was the Attacker, which always seemed to me to have been hastily cobbled together from handy bits of several aeroplanes. I didn't have much to do with it, although I do recall that you had to disconnect the aileron controls to get the guns in or out, which I thought was a desperately bad expedient. Also it was a tailwheel design, which meant that if engine runs were done too close to the edge of the concrete, great holes could be dug by the jet blast. It wasn't with us for long; just as well, I suspect.

Next came the Seahawk, a Hawker design featuring a clever way of maximising the fuselage fuel capacity by splitting the jet pipe so that it exhausted on either side of the fuselage – a feature that reappeared later on the Harriers. Engineering detail on the earliest Seahawks left a lot to be desired but after production was transferred to Armstrong Whitworth things improved greatly. It was a very sleek and attractive aeroplane that pilots said was easy and pleasant to fly. Perhaps a bit too easy and pleasant, leading to over-confidence, because the accident rate was high.

The De Havilland Sea Venom of the same period was an all-weather fighter with two seats and equipped with radar. Its peculiarities were a twin tail boom layout and a wooden fuselage forward of the engine. Those of us who worked on it were prone to remark that De Havilland should have stuck to their origins as furniture makers because tolerances were wide and quality very variable. Every aeroplane was different; you couldn't take a panel from one and fit it to another. Spare panels came with a very generous margin that had to be cut away to fit, which was often a laborious and time-consuming business. My first flight in a jet aircraft was in a Sea Venom Mark 20 and I well remember the incredible smoothness of the engine. The distant vacuum cleaner whine seemed to have no connection to the aeroplane, which was like a magic carpet. The Mark 20 was not equipped with ejection seats and it was generally recognised that attempts to bale out were unlikely to be successful but although I flew in them several times I don't recall this as a matter of serious concern. One took a fatalistic view. When the Marks 21 and 22 came along they had had ejection seats squeezed into them but the limited cockpit space resulted in a rule that no one with a thigh length of more than a certain figure could fly in them because their knees would foul on the canopy frame upon ejection. My thighs were about three-quarters of an inch too long, so that ruled me out

– somewhat to my relief, because Sea Venoms usually flew higher and descended faster than my ears could tolerate. On one occasion the squadron was to go on detachment and I had arranged to travel in a Gannet that unfortunately became unserviceable. My CO declared that I was not to wait for it but to go in a Sea Venom instead. I pointed out how impossible this was owing to the enormous length of my thighs but to no avail. So off we went and, quite illogically, I worried all the way. Being unable to bale out I was able to accept fatalistically, but the thought that I might leave my kneecaps behind filled me with horror. We went at 40,000 feet, too, so there was also a very painful descent and deafness to follow.

The successor to the dreadful Firebrand was the Westland Wyvern. This was a turbo-propellor type with a gas turbine driving two large contra-rotating propellors. In engineering terms it was very well constructed to tight tolerances but it had major engine problems, mainly due to insufficient development, and would have been better with a tricycle undercarriage than the tailwheel layout it had. In many ways it was a very impressive aeroplane that missed being a success.

My personal favourite was the Gannet in its Mark 4 and 6 manifestations; I was never involved with the later AEW Mark 3 version. The Gannet had two engines side-by-side in the nose, both driving into the same reduction gearbox. Two co-axial propellor shafts emerged from this gearbox driving the contra-rotating propellors. The port engine drove the front propellor, the starboard engine the rear one. Either engine could be shut down and the aeroplane flown on one for greater fuel economy. This combined the advantages of twin engines without the undesirable asymmetric effects on one engine that are inherent in the conventional arrangement. From an engineering point of view it was generally well done and easy to look after. The two engines were installed and removed as one unit in a rather clever roll out arrangement. That was just as well because the weak point was the reliability of the Double Mamba engine. Luckily, being double, there was always another engine to bring the aeroplane home.

At about the same time as the Gannet appeared the Naval Staff decided that it might be prudent to have a cheap and cheerful type to do the same job as the Gannet (anti-submarine) from small escort carriers. This was the Shorts Seamew, which I came across at Boscombe Down and flew in on several occasions. On the face of things it should have been a great success because it nowhere approached the limits of technology. It had a single Mamba engine and fixed undercarriage and was generally a very simple aeroplane. Despite all these virtues it was a disaster aerodynamically and to the best of my memory only three were ever built. I well remember one strange aerodynamic phenomenon. The aeroplane was to be flown at its 'never exceed' speed, which was, I think, a little over 300 knots. We were having some difficulty in actually reaching this speed. The aeroplane just didn't want to do it – 295 knots no problem but beyond that it just stopped accelerating. Eventually we succeeded but only in a near vertical dive, howling and shaking at full throttle. It was during one of these dives that I noticed the oddity. The edges of the observer's canopy were fringed with a brush type draught excluder and I became aware that I could see vast tracts of landscape between the hairy extremities of the draught excluder and the cockpit coaming. Not only that but I could also insert my hand in the gap with room to spare. When we levelled out to a more becoming airspeed the gap had closed up firmly and the canopy appeared secure but I knew its nasty secret.

The next generation of naval aircraft, the successors to the Seahawk and Sea Venom, were a giant leap in many ways, an awful lot of which were unwelcome from an engineer's point of view.

Figure 4 The Shorts Seamew

Figure 5 The Fairey Firefly

They were bigger – much bigger – and roughly twice as heavy. They were more complicated and a good deal less reliable. They were a challenge but they were not fun. Life became a lot more earnest. And there were all sorts of knock-on effects; catapults and arresting gear couldn't handle the weight and speeds and suddenly the ships were too small. Various makeshift improvements were made to the flight deck machinery and some real breakthroughs were achieved in deck landing safety but the ships remained too small to the end.

Probably the worst of these new aircraft was the Scimitar. I was never personally involved with this aeroplane and those

Figure 6 The Fairey Gannet

engineers who were, used to congratulate me on my good fortune. The aeroplane was complicated mechanically, which did nothing for reliability, and from my distant viewpoint appeared to be terminally incontinent, exuding every fluid it contained whenever it was at rest. I don't know how the pilots felt about it – they tend to accept philosophically, whatever is provided – but I recall the comment of an American test pilot who was invited to fly one at Boscombe Down. After searching around for something polite to say he ventured, 'Only you British could put so much thrust into an airplane and still keep it subsonic.'

The successor to the Sea Venom was the Sea Vixen. It kept the twin tail boom layout but there the resemblance ended. Apart from being much bigger and twice as heavy, it was complicated, especially so in the fuel department, and stuffed full of electronics. The radar and its associated electronic systems were capable of all manner of marvels when they worked; the trouble was that they rarely did. The root of the problem was contact with the bracing salt sea air – an unhealthy aversion for a naval aircraft. Even I began to feel sorry for our electronic specialists in their warm comfortable workshops for as soon as a malfunctioning piece of equipment was removed from its aeroplane to these balmy surroundings for investigation it started working again. No fault could be found and so it was returned to duty where, with an electronic shudder of revulsion, it died again. Eventually a partial solution was found courtesy of the American space programme and the humid climate of Cape Canaveral, which was causing similar problems with the space rockets. This was Rocket WD40, a magic mixture of fluids but allegedly 70% white spirit. Nowadays most households have a handy can of WD40. It's one of those high technology spin offs, like non-stick saucepans.

Unfortunately no such simple solution appeared for the Sea Vixen's fuel system. The aeroplane had a lot of well-distributed fuel tanks and a complex fuel supply and transfer arrangement using many booster pumps. Not surprisingly, when life in the air got busy mistakes were made and some aircraft and crews were lost. One squadron Engineer Officer devised an ingenious and relatively simple way of improving the system, which was soon incorporated and transformed the situation, but the inherent deficiencies remained and still, occasionally, claimed victims. One day I was called to the site of a Sea Vixen crash near Yeovilton. The crew had had fuel transfer problems resulting in both engines failing on approach to the airfield. The crew ejected safely. The wreck didn't burn so there was plenty to investigate. In the course of the examination of one of the fuel tanks, conveniently opened by the crash impact, the booster pump was found nestling in its proper place. But that wasn't all: there was also a new and unused booster pump in the tank, still wrapped in its sealed plastic bag, to the wonderment of all. As it happened that had nothing to do with the crash.

I always felt that one of the nastier features of the Sea Vixen was the observer's accommodation, aptly known as the 'coal hole'. From this awful position the observer could observe nothing of the outside world since the only transparency was a small square in the region of his right thigh. To this day I have nothing but admiration for those observers who cheerfully occupied that seat day after day and, even worse, night after night. I find it difficult to believe that anyone has fond memories of the Sea Vixen but many of us have not forgotten it because it remained with us for a long time.

Arguably the most successful British naval aircraft was the Blackburn Buccaneer. It originated in a far-sighted and uncompromising Naval staff requirement and an imaginative design. It incorporated some of the latest aerodynamic fashions, such as leading edge boundary layer control, blown flaps and drooped, blown ailerons. It even had a tailplane flap, also blown. All this blowing was accomplished by tapping air from the engines' compressors during take-off and landing when high lift was needed. This enthusiasm for using bleed air extended to electricity generation by means of an air turbo alternator that lived back aft among the liquid oxygen containers and other odds and ends. This was all symptomatic of the besetting sin of British engineering – the urge to be clever at the expense of simplicity. The Gyron Junior engines worked hard to provide the air required for all these services but from time to time it all became too much and the result was compressor stalls and surges of impressive malevolence, with alarmingly loud bangs, often in prolonged series, and sometimes the emission of flames from the wrong end of the engine. The Gyron Junior compressor was a good design but underdeveloped, with too many demands on it, and the Buccaneer Mk 1 was underpowered, especially in the landing configuration. One result of this was that Buccaneer Mk 1 deck landings were conducted with extreme care with very notably steady approaches. The Mk 2 Buccaneer had Rolls Royce Spey engines of considerably greater thrust. The blowing system and its control valves were simplified and the awful air turbo alternator was replaced by engine-driven alternators. The result was a much more user-friendly aeroplane acknowledged to be the world's best of its kind and its time. Sadly the navy was forced to hand these splendid aeroplanes over to the RAF as a result of an ill-considered Defence Review, of which more anon.

The unlamented Sea Vixen was succeeded by an aeroplane designed from the start for naval use, albeit for the US Navy. The Grumman F4 Phantom was, without doubt, one of the greatest naval aircraft of all time, a truly outstanding design and an admirable engineering achievement. It incorporated many of the fashionable aerodynamic advances of the Buccaneer, such as blown flaps, but without the attendant complexity. Its performance was such that despite being designed to operate from ships it was gratefully accepted by air forces, including our own. The Royal Navy had long coveted the Phantom although the quid pro quo imposed by the politicians was that the British version should be powered by a variant of the Rolls Royce Spey. This was a political move intended to maintain British jobs but accepted without argument by the Naval Staff because, on paper, the Spey would provide more thrust with lower fuel consumption. The General Electric engine in the standard Phantom was a well-proven and tough piece of kit with the great merit that all its parts were interchangeable and engines could be built up in the field from spares. These were great engineering advantages but, not unusually, engineering considerations were as nothing in the face of politics and promises of performance, even though the GE engine was also cheaper. Once again we could not resist the temptation to be a little bit too clever. As promises turned into facts and hardware it was found that the Spey engine was a little bit too fat to fit comfortably in the Phantom airframe so the bypass duct was reduced in diameter to fit, with the result that much of the advantages of thrust and fuel consumption were lost. The engine also became temperamental, reluctant to start in cross winds and suffering spectacular compressor stalls for no good reason. In time these defects were cured or at least reduced to acceptable levels. The Royal Navy's Phantoms were still excellent aircraft but we were not to enjoy them for long. Like the Buccaneers they were handed over to the RAF and their carriers were scrapped. The Royal Navy was no longer one of the front rank.

I should not forget the Navy's aviation equivalent of HMS Victory, the famous Fairey Swordfish or 'Stringbag', which was obsolete when the Second World War started yet served the Navy well all through that conflict. The Historic Flight at Yeovilton included one and it came within my orbit for maintenance purposes. This gave me a right to hitch a ride from time to time when a test flight was required. These were gentle affairs and usually undertaken on bright sunny days because the Swordfish was the ideal joy-riding conveyance. The rear cockpit was large and boat-like with low sides and plenty of room for two, especially with the gun mounting removed. The aeroplane was almost permanently fitted with a torpedo but this was a hollow dummy and did nothing to detract from the impression of a benign intent. Starting the engine was a bit of a performance; you needed one strong man or two WRNS to crank away at a handle that turned a flywheel. When the flywheel was up to speed the pilot engaged a clutch that cranked the engine over a few turns and hopefully caused it to start. If throttle and mixture settings were not spot on there would be no start and thinly disguised displeasure on the part of those cranking the handle. Once started the engine would rumble away smoothly and surprisingly quietly. The aeroplane would amble majestically into the air and contrary to expectation gave no impression of fragility, despite the struts and wires and fabric covering. Rather there was a feeling of strength and security and the grandeur of those enormous wings. I can appreciate, however, that things would certainly have felt different with a Focke Wulf 190 on your tail. In peacetime, though, it was sheer joy to fly in it. The view was superb, the speed (about 80 knots) was just right to wave at people on the ground and you could stand up in the cockpit, lean against the wind and look over the pilot's head at the valves on the Pegasus engine bouncing up and down. Once or twice we did dummy torpedo attacks on the dam of the Chew Valley lake. It was all great fun though tempered with great respect for all those who flew Stringbags in action when the days were not necessarily bright and sunny.

It occurs to me that I am a member of a very restricted club of those who have flown behind a Pegasus engine in the Swordfish and in front of one in the Harrier.

10

Helicopters

Soon after I first arrived at the Royal Naval Engineering College we had a visit from one of the Royal Navy's first helicopters. It was a Sikorsky R4B and compared with contemporary normal aeroplanes it was a bit of a throwback. It was made of steel tubes covered with fabric; even the rotor blades were made that way. It looked old-fashioned, curiously complicated and, to me, very interesting. The pilot was persuaded to give some of us a short ride so the first passenger got in and with a great whirling of blades and billows of dust it lifted off to a height of about 50 feet and began to move off. Almost immediately however it came back, landed and shut down. The pilot explained that the cylinder head temperatures were 'going off the clock' because it was a hot day and the passenger was too heavy. We were not impressed by this inability to withstand the rigours of an English summer day and it was to be a long time before I had my first ride in a helicopter. All the same I remained intrigued by the whole idea of helicopters although I knew virtually nothing about them and they did not then feature in our Aircraft Engineering course, which I began nearly three years later. Towards the end of this course we had to prepare a dissertation and it was made clear that this should not be a mere regurgitation of course work but should include original and forward-looking thought. In this regard I had an open and unfortunately blank mind. I did not want to choose any of the subjects picked by my colleagues so in the end, in order to be different, I blithely offered to undertake a thesis on The Control of Helicopters. I did have some misgivings but in my innocence ignored them.

It was not long before it became clear that I had made a serious mistake and was completely out of my depth. There was nothing on the subject in the College library. In fact at that time there wasn't very much on the subject anywhere but the College sent off for a publication that duly arrived and was eagerly seized. The book completed my intellectual humiliation. It held page after incomprehensible page of mathematics with only very occasional outbreaks of English prose. I am sure that in mathematical terms it treated the subject exhaustively but it did so without mentioning precession or coriolis, for instance, in the English language and so left me ill-prepared to treat the subject either originally or forward-looking or indeed at all. It was too late to change so I had to press ahead as best I could and eventually I did produce a thesis of sorts. Rather surprisingly it attracted no adverse comments (or favourable ones) because from end to end it was largely waffle.

I put the whole thing behind me and forgot about it but it came back to haunt me when, a few months later, I joined HMS Glory. Back in those distant days every aircraft carrier was accompanied by an attendant destroyer that would take station off the port quarter while flying was going on with the objective of rescuing aircrew when their aircraft fell into the sea. (Note when, not if.) However, Glory had been selected to operate without the company of an expensive destroyer. For the first time in the RN the ship would be equipped with her own rescue helicopters. These were two Sikorsky S51 Dragonflies and they arrived on board only a week or two after I did.

Officially I was the Air Ordnance Engineer Officer but my boss, checking up on my history, discovered the subject of my thesis and acted accordingly. He was very suspicious of these unnatural aeronautical devices and wanted as little as possible to do with them. You are the expert, he said, so those helicopters are yours, all yours. In the next few weeks I learned a great deal about helicopters that was not written in books about them. For a start their ability to land and take off from any small site is a very mixed blessing. When an aircraft carrier comes into harbour the engineers can get on with servicing their aeroplanes except, that is, for helicopters. In particular, my two helicopters were a novelty and when we were in harbour every senior officer of any consequence felt entitled to use them as personal ship-to-shore taxis. Only darkness brought a welcome respite but the result was that most of our maintenance was done in the small hours and the same crews had to look after the helicopters during the day. There was a great deal of maintenance, too, because those early helicopters were none too reliable. Rotor heads required frequent greasing that all too often revealed a failed flapping or drag hinge bearing. I learned to be very wary when a pilot remarked, approvingly, that the helicopter was flying smoother because that was often the precursor of a flapping hinge failure. An early example of a modern semi-rigid rotor perhaps?

The Dragonflies had arrived with maintenance manuals covering daily inspections and the like but we were still waiting for the more in-depth maintenance manuals when one of them landed making an unusual noise and a pilot complaining of loss of power. We soon discovered that the cause was a spark plug that had blown out of a cylinder head, leaving an ugly and irregular hole. The engine was an Alvis Leonides of, I think, 550 horse power; beyond that we had little information other than how to top up the oil. The Leonides cylinders and heads were integral so the whole cylinder had to be changed. Fortunately we had a spare and we set to work to change the cylinder. I forget which one it was – let's say number 7 – and it didn't look too difficult. In fact it took a fair amount of cunning manoeuvring to get the cylinder off as there was not a lot of room. Getting the spare one on was another thing entirely. There *had* to be a way but we could not find it. Eventually, after a night's work someone did something different – we never knew what – and it slipped into place. In the morning the helicopter was airborne bright and early as usual. A few weeks later we received the remaining maintenance manuals and noted that 'all the cylinders can be changed with the engine in situ except No 7 for which the engine must be removed'. We knew different – but not how.

Apart from their function as senior officers' taxis when in harbour the helicopters also offered luxury ship-to-ship transportation when at sea. No more the tiresome business of lowering a seaboat for a rough and damp passage followed by an undignified leap for a ladder. As a consequence inter-ship socialising became more popular among the governing classes. This was all very well when the other ship was an aircraft carrier but quite another if it was a cruiser. In that case I suffered considerable foreboding. In a cruiser the only available landing area was the quarterdeck – a holy place where you had to salute on entering that was maintained in a state of excessive shiny cleanliness. In particular the deck itself was teak scrubbed to an improbable level of near whiteness, the pride and joy of the 1st Lieutenant and the Executive Officer, who jointly claimed this territory.

Now the Alvis Leonides engine was, on the whole, no great trouble but one of its little idiosyncrasies derived from its crank-case breather system which, from time to time and entirely unpredictably, dumped a pint or so of blackened burnt oil on to an unsuspecting world. That was of no great consequence when over the sea or even over the much-used steel flight deck of an aircraft carrier, whence it could be easily wiped up. Over a teak quarter-deck it was cataclysmic. Signals flew, angry signals, culprits were sought and blame was assigned. Usually it was assigned to me, if not as an act of deliberate sabotage at least as culpable negligence in not preventing the occurrence. My suggestion of some kind of improvised external nappy was not well received and the explanation that the event was entirely random was simply not believed.

Looking after those helicopters was an educational experience but not one I was anxious to repeat. It was a constant anxiety with only the two of them. We never failed to have a plane guard helicopter hovering off the port quarter during launch and recovery but sometimes it was a near thing.

That was my only hands-on engineering contact with helicopters despite a lasting interest in them and despite being actually appointed to an anti-submarine helicopter squadron as the Air Engineer Officer. I have long forgotten the squadron's number but it was equipped with Whirlwind Mk 7s and I did not join the squadron because they all fell into the sea and the squadron was disbanded. A lucky escape. The reason for this unfortunate performance was the Whirlwind's engine, which was a big brother of the Alvis Leonides called the Leonides Major. This engine developed the distressing habit of stopping suddenly, usually when the helicopter was hovering low over the sea. The cause was obscure and baffled the best brains for long enough to make its eventual discovery of only academic interest.

11

And Another Thing

HMS Gamecock was not a ship. It was a Naval Air Station but not a normal one. It was a grass airfield and so not suitable for jet aircraft and it was as far from the sea as it is possible to get in this island. None of these things mattered very much because its purpose was to train Naval air mechanics. It was hidden away close to Nuneaton and impinged little on the front line Navy. It certainly impinged little on me until I was appointed there. The appointment came as a shock to me because I did not consider myself training material. My manifold deficiencies had been pointed out to me during my years under training and I was clearly no role model for the young and impressionable. So I had concluded, with relief, that I would never have to do a training job.

It seemed that the appointment was not an error and it got worse. I would not be involved in the engineering training of these young recruits but in the formation and guidance of their characters. I duly joined feeling enormously embarrassed and fraudulent. In retrospect the following two years seem surreal, something that I observed with disbelief from a distance. When I joined it was pointed out to me that my job was especially important and would demand total dedication. 'Are you dedicated?' I was asked. The question was an embarrassment to me because, for one thing, I was not sure what was meant by it and also because it was impertinent, whatever it meant, so I took refuge in a flippant, dismissive reply. The question recurred from time to time and it was not the least of the impertinences. It appeared that my deficiencies, which had for years been unremarked, had again become obtrusive and demanding of immediate correction. I did my ineffective best with the hair, I practised my salute until it was correct most of the time and I went to church quite often to set an example to my trainees. Needless to say this was an example ignored by most of them.

The business of church going or not was a consistent source of complaint from my superiors and a recurring irritation to me. I thought that it was an unjustified intrusion into the remains of my private life and irrelevant to my job.

Although the self-righteous piety of Gamecock's wardroom was hard to bear, there were compensations in the job itself. They stemmed mainly from the trainees who were, with very few exceptions, good-natured and well-motivated young men who blossomed visibly over the duration of their training, both physically and in self-confidence. It was a pleasure to teach them and we must have been getting something right despite my dubious dedication. Preparing the training programmes was always a disputatious process. We were preparing these lads for the job of maintaining aircraft but there was a strong element on the staff who treated technical lessons as if they were exposure to some insidious virus requiring the immediate application of an antidote. Suitable antidotes were parade training, physical training or seamanship. Any suggestion of two successive periods of technical subjects was dismissed with horror. At the same time there was a reckless inclination to experiment, without adequate forethought, in a manner that would never be tolerated if the subject was machinery. Fortunately the subjects were these adaptable young men who produced much the same results whichever bright idea was currently in favour. In the years since I have noticed that free and easy experiment is endemic in the world of education. I think it may be a virus.

In one way and another the Navy gave me some considerable variety of engineering experience much of which was routine, sometimes boring and frequently uncomfortable. Some would call it character forming but it added up in time to a perception that what I knew about naval engineering was worth knowing and that the opinions I had formed around this knowledge and experience were not to be lightly dismissed even when they strayed beyond my own specialisation. Like invention, this perception was usually best kept to oneself but from time to time I allowed myself to forget that and it was usually a mistake. However, all the foregoing may help in understanding some of the otherwise incomprehensible influences and motivations behind the often chaotic activities surrounding the next theme – the genesis of the Ski Jump.

12

A Seed is Sown

The process of invention can be seen coherently only with hindsight. Milestones and turning points are then often clearly identifiable, although at the time they are all too frequently invisible. Hindsight also brings order into what was, in fact, chaos and turns it into a logical progression. Bearing this in mind it is probably the events of a weekend in 1955 that marked the threshold of the Runway in the Sky. I was a trials officer at the Aeroplane and Armament Experimental Establishment at Boscombe Down. This establishment carried out all the practical trials on new military aircraft and their equipment to prove that they were suitable for service use.

I was the duty officer on this particular weekend when a lorry arrived bearing a large triangular packing case. When it was unloaded the driver required a signature from me to certify its safe delivery. He said that there was an aeroplane inside but I doubted that: part of an aeroplane – most likely a tailplane in view of its shape – was my opinion. However, when unpacked on the Monday there, sure enough, was a quaint little aeroplane. It was called the Short SC1. It had no fewer than five small jet engines; one lay horizontally in the tail, the other four were clustered vertically amidships. It was intended that it would take off and land vertically, with no ground run, using the thrust of the four vertically mounted engines. When comfortably airborne the single engine in the tail would accelerate it horizontally until it reached a speed at which its seemingly inadequate little delta wing could support its weight and allow the four lift engines to be shut down. To land, the procedure was to be reversed; start up the four lift engines and, as you slow down, open them up until they are taking the weight and you can descend vertically into a small space. QED.

This was an intriguing concept. It held out the prospect that a small, high-performance aircraft could be operated from a small ship such as, for instance, a frigate. Hitherto this was strictly helicopter territory. However, the prospect was a very distant one because the sad fact was that the SC1, taking off vertically, would struggle to carry sufficient fuel to take it out of sight. Its great drawback, as I saw it, was that in normal flight it was lugging around four engines that were so much useless weight and volume until the time came to land. There was a school of thought that claimed that this arrangement would be the most efficient for future vertical take-off aircraft but my own instincts were against it. All the same it was a timely catalyst for thought because these were exciting times on the flight deck. The last of the high-performance piston-engined aircraft were being replaced by the first jet and turbo-propeller aircraft and it was already becoming evident that the size, weight and speed of the new aircraft were pushing the capability of our aircraft carriers to their limits. Accidents were almost a daily event and casualties, sadly, were not unusual. The idea of being able to take off and land very slowly or even at zero speed had an obvious appeal.

Despite the appeal of the concept there could be no argument. The SC1 was not the right aeroplane; nor were any of the other vertical take-off experimental aircraft that appeared at around this time. What they had in common was insignificant payloads, inadequate range and endurance and considerable deficiencies in controllability. As for my thoughts on operating aeroplanes from frigates, it was as well that I hadn't aired them in public because I still had a career to think of and – as I was to discover – such thoughts were heresy.

13

The Right Aeroplane – and a Wrong One

At Kingston on Thames a small but talented team at Hawker Aircraft (later to become British Aerospace) had designed an aeroplane called the P1127, which appeared to have none of the deficiencies of the Short SC1. It had only one engine that performed all the vertical and horizontal thrusting functions required by a simple and cunning arrangement of swivelling nozzles – no hauling of spare engines around the sky. To me it was brilliant. This was the right aeroplane and perhaps my wild speculations about operating from frigates were not beyond possibility. Unknown to me, the P1127 had peculiar problems of its own but in any case I had a job to get on with and it had nothing to do with V/STOL aeroplanes. My proper job was as the Deputy Development Project Officer for the Buccaneer and, as has been mentioned, this was giving me plenty to think and worry about. All the same, in recesses in the back of my mind, naughty little thoughts concerning vertical take-off and small ships would peep shyly through from time to time.

Meanwhile, in an office not far from mine, the P1127 was metamorphosing on paper into something bigger but definitely not better. They called this paper monster the P1154. The main cause of its monstrosity was that it was to be a joint service aircraft. This joint service concept is a logical and fatally flawed conclusion of the tidy methodical mind. Despite all the disasters and the millions of pounds lost in consequence the tidy minds keep returning to the same conclusion. To be fair they are also motivated by an urge to quell controversy. In 1918 the Army and the Navy were quarrelling over the supply of aircraft and creating a great deal of untidiness. The solution was to take away their aeroplanes and give them to an entirely new organisation, which created an entirely new quarrel and made sure that neither the Army nor the Navy got the sort of aeroplanes they needed. Pursuing this persistent line of thought, the conclusion in 1961 was the P1154, which, in a delicious irony, neither of the prospective customers wanted. The RAF wanted a short-range, single-seat interceptor and the Navy wanted a long-range, two-seat, all-weather fighter. In fact the Navy knew exactly what it wanted and where to get it. They wanted the F4 Phantom as supplied to the US Navy.

In the end the technical incompatibility of the two P1154 requirements was too obvious to be denied even by those who had instigated the project and it was allowed to die, to almost universal relief. The Navy got its Phantoms.

In the Buccaneer office we had watched the P1154 battle with interest and not a little *Schadenfreude*. One aspect of the controversy became clear: even if the P1154 had met the Navy's requirements in every detail the Navy hierarchy would have fought like tigers to reject it because the P1154 did not necessarily need large aircraft carriers from which to operate. Admirals want big ships – in fact, they *need* big ships to accommodate them and their staffs at sea – and there is a long tradition that British admirals fight at sea, not behind a desk on land. I was aware of this but paid insufficient attention to it, being totally convinced of the advantages, as I saw them, of being able to use smaller ships as aircraft carriers.

This is probably as good a point as any other to set out these presumed advantages.

We start from the presumption that Great Britain is now a small nation that is unable or unwilling to afford the kind of world-class navy to which it has hitherto been accustomed. If, as a nation, we insist on possessing large aircraft carriers then our financial constraints will limit their numbers. At the time of writing we think we can afford two, though that is by no means generally agreed.

Now an aircraft carrier is a warship and as such must go in harm's way where there is a possibility that it will be disabled or sunk. The outcome of subsequent events will depend, to a large extent, on whether such disabling or sinking represents a mortal blow.

By way of example, consider Admiral Beatty at Jutland. He was faced, in dauntingly quick succession, with the destruction of two of his force of battle cruisers. This gave him concern, forcibly expressed, for the quality of his ships but not for the outcome, because the fleet could afford to lose them – there were six remaining.

Twenty-five years later the loss of the battleships Prince of Wales and Repulse was a decisive victory for the Japanese. It was also a severe blow to morale because the battleships were also symbols of national power. But in those days we did have a world-class Navy that was able to survive those losses and the undoubtedly serious wound to our morale. This is no longer the case and in 1982 the Prime Minister herself acknowledged that if either of the two carriers involved in the recovery of the Falkland Islands had been sunk or disabled then the enterprise would have had to be abandoned.

However, when I came to this view all that lay in the future. What seemed clear to me was that the aircraft carrier had become too precious to be risked and therefore was no longer an asset but a liability that would tie up too much effort in providing its protection. It was, in other words, a Naval dinosaur.

The solution, it seemed to me, was to put the aircraft into numbers of smaller, more affordable warships and into converted merchantmen. After all it was the escort carriers that were decisive in winning the Battle of the Atlantic in the Second World War and now, with the advent of V/STOL aircraft, we were in sight of the means to do it. We had a prospect of organic naval air power that could be concentrated or dispersed as necessary and that we could afford to put at risk.

I kept this view to myself not from fear of ridicule or of damaging my career but because the subject never arose. And in any case at this stage it was not a practical reality. It was part of my hobby of invention – a fascinating subject but a private one.

Meanwhile with the demise of the wrong aeroplane (the P1154), the P1127, – the right aeroplane – was evolving into something even better: the Harrier. This was going on with little enthusiasm on the part of the RAF and none at all from the Navy. In fact the whole project had little official support from any quarter except, surprisingly, the United States, which was helping with funds.

14

The Harrier – Star of the Show

I have mentioned before the debt I owe to luck and the Harrier was yet another manifestation of the lady. Just when the right aeroplane was needed, along it came. Probably the main reason why it was the right aeroplane – in fact the only candidate, all others having failed – was that it had escaped the clutches of the Service Departments and the Ministry of Supply. It was, effectively, a private venture, like the Spitfire before it, until officialdom, seeing its success, leapt on the bandwagon. As a consequence it evolved from the P1127 in the hands of a team that had a clear objective in mind and the freedom to ignore the siren calls to divert from that objective. There was nothing 'nice to have' about the Harrier. What it had was necessary and uncompromised. One of the things it had was a system of thrust vector control whereby *all* the engine thrust could have its direction varied from horizontal (pointing backwards) downwards through an angle of 110 degrees (that is, 20 degrees forward of vertical), virtually instantaneously, via a lever in the pilot's left hand. Furthermore, at all thrust angles the thrust line remains passing through the centre of gravity of the aeroplane. To this day no other V/STOL aircraft has this capability and so cannot take full advantage of the Ski Jump.

Another thing it had was relative simplicity. Years ago an aviation sage coined the phrase 'simplicate and add more lightness'. More recently, we had KISS (Keep It Simple, Stupid). Both are often quoted self-righteously in design circles while being ignored in the interests of expediency. They were not ignored in the Harrier and one consequence is a reduction in the number of those things with a distressing tendency to go wrong. One of the impressive features of the Ski Jump trials was the way the aeroplane turned up serviceable morning after morning.

Of course uncompromising simplicity can have its downside and in the case of the Harrier some desirable design for maintenance aspects have had to be sacrificed, most notably those concerning the engine, which sits in the centre of the aeroplane. It is absolutely essential that it should do so but it has the unfortunate result that it becomes necessary to remove the entire wing in order to do an engine change. Furthermore the engine is a very tight fit in the fuselage, leaving little room for error when lifting it in or out. This makes engineering life quite nerve-wracking when the job has to be done in the hangar of an aircraft carrier at sea and pitching and rolling in a disobliging manner.

This is not the place for a detailed technical description of the Harrier but it is as well to be aware of some of the peculiar characteristics of the aeroplane that are not obvious and that are relevant to the objective I had in mind. John Farley, Harrier chief test pilot, summed it up quite succinctly. He said that it was an easy aeroplane to fly and an easy aeroplane to get into trouble in.

First a few facts of V/STOL life.

Typically, the thrust of the Harrier's Pegasus engine is 21,500lbs from which the uninitiated might conclude that the aeroplane should be able to take off vertically at a weight of close to 21,000lbs on a good day – but it is not so. Apart from the fact that you need a bit of thrust in reserve for manoeuvring, nature imposes another of her irritating penalties. In this case it is the downward flow of entrained air induced by the four jet nozzles. This mass of air impinges on the top of the wing and fuselage and causes a *downward* force opposing the engine thrust. This force is far from negligible and means that the maximum vertical take-off weight is, more realistically, something like 18,500lbs.

The maximum weight of the aeroplane is, say, 25,000lbs but in order to take off at that weight the aeroplane needs some aerodynamic lift, which it gains by accelerating along a runway with jet nozzles pointing aft until, at a predetermined airspeed, the nozzles are directed downwards at an angle of, typically, 50 degrees. The combination of wing lift plus jet thrust is greater than the weight of the aeroplane, which obligingly takes off.

This method is called a short take-off (STO). In fact it is not very short, being at least 400 yards, but it is short compared to a conventional aircraft take-off where the distance moves into thousands of yards. That is the theory. In practice more of nature's complications enter the story. The theory holds good while the aeroplane is accelerating with nozzles aft but at the magic moment when the nozzles are rotated downwards through 50 degrees the jet blast, striking the ground, ricochets up again and strikes the underside of the tailplane applying a powerful upward force to it. The effect of this is to pitch the aeroplane nose down so that it tries to run along on the nose wheel (known as wheelbarrowing). This also has the effect of reducing the angle of attack of the wing and consequently the aerodynamic lift. In order to overcome this effect the pilot has to apply considerable nose up control. The aeroplane then takes off but as it clears the ground the jet blast no longer strikes the tailplane and the pilot has to reverse the nose up control. All this must be done at a time when control power is limited and there is little room for error. Some of the more tricky aspects of flying the Harrier could, perhaps, have been eliminated or eased by electronic artificial stabilisation but in those days electronics were the biggest contributors to military aircraft unserviceability and John Fozard, Harrier chief designer, very wisely in my view avoided them wherever possible. As we shall see, the Ski Jump eliminated many of these piloting problems but I cannot claim that this was among my intentions because at this stage I knew very little about the Harrier beyond the very limited published information. In any case there were other things occupying my attention.

15

Revelations East of Suez

I had very little spare time to devote to the Harrier and small aircraft carriers because more immediate affairs were pressing upon me and they were not welcome. I had had nearly three years with the Buccaneer and had come to know it well. It seemed likely that I might be the Air Engineer Officer of the first squadron on formation – not certain but a logical outcome. However that was not to be. 801 Squadron formed and went to HMS Victorious without me. I didn't have to wait long wondering what my fate was to be. I would be following the Buccaneers to Victorious but with a more tenuous connection. I was appointed as the Flight Deck Engineer Officer (FDEO) in charge of the flight deck machinery, that is catapults, arresting gear, aircraft lifts, liquid oxygen plant, etc. This was all marine engineering and I had had no contact with it for ten years or more. My Buccaneer expertise was wasted and I was not happy.

There's a saying in the Navy, 'Grumble you may but go you must'. There was nothing to be done about it but to accept the situation but it must be admitted that it was with misery that I flew off to Singapore leaving an equally unhappy wife and two small children for what we thought would be two years.

HMS Victorious had been launched during the Second World War and since then had been subjected to many piecemeal additions and modernisations. She had just emerged from the latest modernisation, which had kept the ship in dockyard hands for years. British naval dockyards were, in general, not good for ships. In Pepys' time they were notorious for bad workmanship and corruption; more recently they had become very highly unionised, which had much the same result. What I found in Victorious was not inspiring. The modernising of my department included uprated catapults and arresting gear to handle heavier aircraft like the Buccaneer and the installation of two liquid oxygen production plants. The catapult machinery was a particularly poor piece of work. Granted the space available was barely adequate. In the catapult machinery rooms under the flight deck I could not stand upright – the headroom was about the same as Nelson's Victory. But that was the least of the problems. The space was dominated by two large steam-driven hydraulic pumps that, it was said, had been built in 1912 to provide hydraulic power to the 15-inch turrets of the new battle cruisers. Being steam-driven with the usual steam leaks there was a lot of wild heat, aggravated by a tropical sun on the plates of the flight deck immediately above.

The hydraulic system itself was a bit of a nightmare: apparently unplanned, with pipes having been run on a first come, first served basis and mostly unsupported, so that they whipped and banged against each other and surrounding structure as fluid moved through them. They were all too often immersed in an unsavoury mixture of rusty water and hydraulic fluid from the many leaks against which we fought a running battle. The pipe joints were held by bolted flanges more appropriate to domestic plumbing than to high-pressure hydraulics. The whipping of the pipes and the impacts resulted in leaks and fractures that could put the catapults out of action and at the time when I joined the ship this was occurring on a daily basis. Our non-engineering masters tried to be tolerant about this but their patience wore thin.

There is an unspoken tradition in the services that you do not waste time and energy wailing about the quality of your

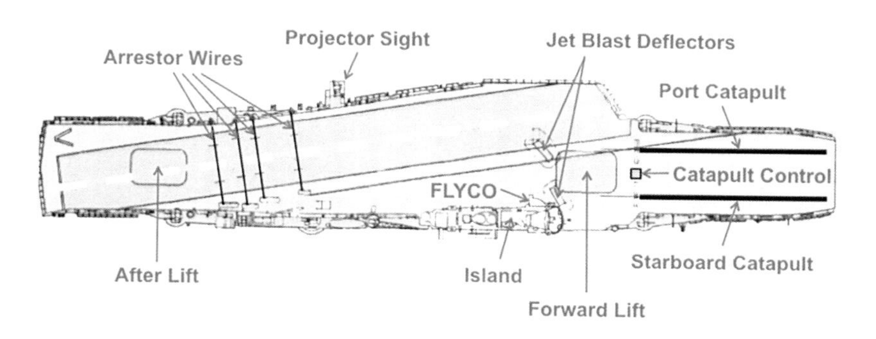

Figure 7 The flight deck of HMS Victorious – plan view

equipment; you put up with it and get on with it. We put up with it and, as opportunities occurred, re-ran the pipes and replaced the flanged joints with brazed joints so that things gradually got better, though not as good as they should have been. I had a first-class team and we began to take a perverse pride in our achievement while remaining strongly of the opinion that it should not have been necessary.

I have dwelt on this because this situation was absolutely typical of the time and was accepted at almost every level because we were used to it. It was not at all unusual for ships to emerge from the dockyard after months or years of refitting with multiple serious defects that the ship's staff had to make good at sea. Shortfalls in quality were due, we were told, to lack of money, an excuse that, on the whole, was accepted. My eyes, at least, were about to be opened.

I had had, up to this time, only limited and sporadic contact with the US Navy. Like most British naval officers I grudgingly accepted that it was the biggest navy in the world but truly believed that the Royal Navy was the world's best and the most efficient.

Figure 8 HMS Victorious – general view of flight deck

Figure 9 HMS Victorious, forward end of the flight deck – catapults and control position

A day came when we were to operate with the USS Kittyhawk, during which we would launch and recover each other's aircraft. So I was sent by air in advance to Kittyhawk where I spent three days finding out the peculiarities of their aircraft and equipment. I found myself being surprised and impressed in ways that I had not anticipated. The USS Kittyhawk was about twice the size of Victorious but, that aside, it was another world in a technical sense. Not, as one might have expected, in terms of advanced technology, but in simplicity and sensible practicality. Her catapults were very similar to ours but longer (250 feet compared with our 150 feet) and she had four of them. But the real eye-opener was the catapult machinery rooms. These were cool quiet places where electrically driven hydraulic pumps hummed contentedly and, amazingly, the deck under the walkway gratings was painted an immaculate white. I couldn't help commenting. 'Oh, yes,' said my guide, 'that's so that any leaks will show up.' 'You get many leaks, then?' I asked, rather hoping that these people had some weaknesses. 'Nope,' he said, 'but when we do we sure want to know about it.' It turned out that my American equivalent managed with about half the personnel I had despite having nearly twice as much machinery to look after. The simple reason for that was that his machinery was much more reliable than mine – not more expensive – and it was more reliable because the US Navy wouldn't accept less. I flew back to Victorious in a humbler frame of mind.

Victorious was typical of British fleet carriers of the 1960s period. Her flight deck layout is shown in Figures 7, 8 and 9. Aircraft to be launched were positioned in a herringbone formation down aft among the arrestor wires waiting to taxi forward in turn to form an orderly queue behind the two catapults to which they were loaded alternately. Once on the catapult the jet blast deflector (JBD) was raised to divert the jet blast overboard. When the aircraft's engines had been checked at full power it was launched on a signal from the duty Flight Deck Officer. On launch the JBD lowered automatically to allow the next aircraft to be loaded. From time to time the JBD would stick in the up position due to distortion and seizure of the hydraulic jacks. When this happened the catapult crew had to go through a manual lowering routine, which took a minute or two and held up the launch to displeasure all round. If all went well the idea was that the last launch would be from the starboard catapult so that landings along the angled deck could start the moment the port catapult was clear and its JBD no longer causing an obstruction.

When landing, the aircraft would catch one of the arrestor wires with its arrestor hook and would be brought to rest from a speed of 100 knots or more over the deck in a distance of 200 feet. The hook was then disengaged and the aircraft taxied forward to park in a herringbone formation on either side in the catapults area. As soon as the last aircraft landed those that were unserviceable were taken down to the hangar on the lifts, replacements were brought up and all aircraft repositioned for refuelling and rearming ready for the next launch. Usually this was 75 minutes after the previous launch so there was no time to spare.

To see a well-drilled flight deck in action was sheer poetry – a kind of ballet of men and machines combined with high drama and ear-splitting noise. Over all there is the continual roar and buffet of a 30-knot or more wind over the deck as a background. It is all very tiring and, as has been described, no place for the unwary.

We were a well-drilled flight deck; we were good and we knew it. All the same it does not take much to turn the poetry into ugly chaos because there are too many sensitive points. A glance at Figures 7, 8 and 9 will show some of them. For instance, if the after lift gets stuck in the down position, aircraft can't land. If the forward lift sticks down, you can't launch. If a JBD fails to lower after a launch you can't load that catapult and if it happens to be the port JBD you can't land either, because it obstructs the angled deck. If an aeroplane becomes unserviceable on either catapult it holds up the launch (and the landing) while you get it out of the way. If either catapult becomes unserviceable the launch rate is obviously halved and at least one aeroplane will have to be repositioned. These problems are not an unavoidable part of carrier operations but they are a natural consequence of using a ship that is really too small for the type of aircraft embarked.

Our machinery defects were too often due to poor design. Most of them could be put right in an hour or two or even in a few minutes but the design deficiencies that lay behind them could not always be corrected so that they were bound to recur time and again – occasionally in diabolical combination.

For example: the last launch was a Buccaneer from the port catapult and he was having some problem with his instruments, causing a prolonged period at full power. This was unwelcome as another Buccaneer had declared an emergency and requested an immediate landing. At last the tardy Buccaneer was launched to all-round relief but the relief was short-lived because the JBD remained up instead of lowering automatically, thus partially blocking the angled deck. At the same time thick grey smoke came billowing from around the catapult and the edges of the forward lift. 'Fire, fire, fire' came the broadcast but where was it? Lots of smoke but no sign of its origin. It was decided to lower the lift a few feet to see if the fire was in the lift well or under the flight deck. So down went the lift about four feet and it was clear that wherever the fire was it was not there. Up lift was the next order and at this point things got really complicated because the lift would not move. The situation now was that we had an airborne emergency, a fire of unknown origin, a JBD stuck up and a lift stuck down. Since most of this was in my territory my headphones were loud with advice and exhortation. In order to think I unplugged the headphones and tried to figure out some common cause. A phone call to the lift machinery space brought no response – there should have been someone there – events were getting beyond me, I couldn't think straight or see properly in the smoke, which was stinging my eyes. In sheer frustration I kicked an electrical junction box that was conveniently positioned on the wall of the lift well. And the lift came up.

To me this was something of a miracle because amid all the doubt and confusion one thing I knew for certain – that that particular junction box had no connection with catapults, lifts or JBDs. I looked around, the lift was up, my splendid catapult crew had lowered the JBD, the smoke was thinning and back aft a Buccaneer had landed and was taxying forward. All was well but what on earth had been happening? It had started with the Buccaneer on the catapult. They were loaded in a tail down attitude so while he did his prolonged full power checks the deck was subjected to very hot jet blast at close quarters. Now the deck at this position consisted of two layers of steel plate about eight inches apart with the intervening space filled with rock wool. Over the years this rock wool had become impregnated with grease and jet fuel that became ignited by the heat and smouldered, creating the spectacular smoke.

The JBD problem was the recurring one mentioned earlier.

The lift had stuck because its hydraulic pump motor had overheated and tripped – a not unknown event.

The Engineering Mechanic in the lift machinery space was not answering the phone because he was busy starting up the stand-by pump.

All these events were random and entirely unconnected. They merely happened simultaneously and overwhelmed my thought processes.

However, my mental paralysis had some advantageous effect in that at least one person had mistaken it for calmness particularly with regard to the rocket. The rocket?

Apparently the landing Buccaneer was carrying a misfired rocket that detached itself on landing and came skipping up the deck a few feet from me. Of course I didn't see it owing to the smoke and my streaming eyes. I saw no need to mention that.

While this sort of thing was exasperating and adversely affected our operational efficiency most of the manifestations of poor design and workmanship were not lethal. But not all.

The Buccaneer taxied up to the port catapult and was loaded to it. The holdback and the launching strop were attached and the shuttle manoeuvred forward to tension everything. The pilot opened the throttles full in preparation for the launch. While all this was going on the Petty Officer in the catapult control position was charging up the pressure in the steam accumulator to the value required for that Buccaneer. It seemed to be taking rather longer than usual: normally it was charged up and ready well before the loading process was finished. Still, it was ready soon enough not to delay the launch and raised no alarm signals to the Petty Officer or to myself, watching. Catapult ready, pilot ready, the Flight Deck Officer dropped his green flag and the Petty Officer pressed his catapult firing button. The launch valve opened and, to my horror the steam pressure fell right away to a pressure far below what it should have been. I knew that a disaster was about to follow and two seconds later the end speed indicator confirmed the opinion. The launch speed was 20 knots slow. As it cleared the end of the flight deck the Buccaneer fell below my line of sight and I was sure that the next thing to be seen would be an enormous splash. When it didn't appear I dared to hope and ran forward where, to delight all round, the Buccaneer could be seen flying. Barely and very low, the wing tip vortices and jet blast were leaving trails in the sea, but flying. Our technical investigation disclosed disgraceful standards of quality control, supervision and workmanship. The failure is explained in Appendix 1 but in summary a blanking plate on a vital steam pipe had been left held in place by four casual tack welds instead of the 360-degree Grade A circumferential weld specified.

The background to all this was the Cold War between NATO and the Soviet Union which hung over us and threatened at any time to become a real war. Mercifully it never did but in the Far East we had an additional Cold War, or 'confrontation' as it was called, between Indonesia and the newly independent Malaysia. Our job was to protect the Malaysians. Confrontation got a bit warm at times on land and shots were fired but things were pretty quiet at sea. In about September 1964 Victorious went south to

Perth, Western Australia for 'rest and recreation'. There wasn't a lot of rest because of the marvellous Australian hospitality and in due course we dragged ourselves back to sea and headed north. Our route back to Singapore took us among the Indonesian islands. We had two choices, through the Sunda Strait or the Lombok Strait. Either route took us very close to shore with very little room to manoeuvre. The Indonesians claimed that they would not allow us to use either passage and issued blood-curdling threats should we attempt to do so.

We chose the Lombok Strait. We were not worried by the Indonesian threats, nor on the other hand were we complacent. They might have the nerve to attack us and they did have some modern Russian equipment although we doubted if they were trained in its use. So one morning found us approaching the Lombok Strait with the ship at action stations, steam on the catapults and a Sea Vixen at immediate readiness loaded on each. Further aft, armed Buccaneers stood ready at five minutes' notice.

The morning wore on and by the time we were nearing the narrowest part of the strait there had been no significant response from the Indonesians. There was a slow fly past by a solitary Gannet and a submarine had surfaced in the middle distance but these were gestures and not at all aggressive. We were free to admire the beautiful islands passing on either side. It was hot, the wind was from aft and exactly matched the ship's speed so that there was no cooling breeze over the deck and the sun was fierce. After a while the heat and the glare became almost unbearable and my feet were killing me. I went into the island where there was a water cooler and chatted to the Aircraft Control Officer while I rested my poor feet. After a while I went back out on deck where the glare blinded me momentarily and, as sight returned, an uneasy feeling that something was not right began to bother me. Then, with sinking heart, I spotted it. The slots in the deck in which the catapult shuttles ran, normally about two inches wide were down to one inch and the shuttles were clamped immovably. Both catapults were out of action. I realised the cause, gave the necessary orders and reported the news. As expected I was then invited to the compass platform to explain matters to the Captain. Would I be court martialled for negligence or merely logged and removed? In the circumstances the Captain's reaction was extremely kind: 'Bad luck, Doug. Fix it as quick as you can. Radar shows a lot of air activity.'

Fixing it was easy. We hosed down the deck between the catapults so that it contracted with alarming creaks and crackings as it cooled and in about ten minutes we were back in business.

The cause was lack of cooling wind over the deck, hot sun and continuous steam on the catapults, an unusual combination that made the deck expand to an unprecedented extent thus closing the catapult slots. We would have been unable to launch and helpless – a sitting duck – while our main armament, those two double-barrelled 18-inch steam cannon, were out of action. It was only for ten minutes but it could have been a crucial ten minutes (see Appendix 2). As it happened, nothing happened and we returned safely to Singapore.

For me, though, it was a moment of truth and the more I thought about it the stronger became the mixture of my feelings about the incident. They were mainly a combination of shame and anger plus frustration. Shame because of the state to which the once proud service I belonged to had allowed itself to be reduced. We were making inadequate attempts to keep up with the American Joneses and looking pathetic as a result. Shame because of the state of British engineering in our ships and shame that we engineers had allowed ourselves to connive in this decline. Anger at ourselves for accepting our lot without protest and at our political and service masters for their deceptions over the years.

One thing was clear to me: Victorious was fit only to fight a foe who was unlikely to fight back. As a modern warship she was a liability rather than an asset.

The frustration was due to my complete inability to see any practical way of doing anything effective about it. Conversations on the subject with colleagues had to be undertaken with caution and it soon became clear that most accepted the situation and had never even thought of questioning the status quo. I appeared to be alone in my conclusions.

16

Bank Block

What turned out to be my last seagoing appointment came to an end and I left Victorious without regret. My next appointment was in the Naval Aircraft Department in Bank Block of the Old Admiralty Building. I was very glad to be with my family again but in view of my developing opinions regarding operation of aircraft from smaller ships the job itself was one of life's little jokes. In bank block we were busy preparing the specifications for a new, bigger and even more expensive fleet carrier. It didn't yet have a name and was known by its designation – CVA-01. This ship would have been properly equipped to handle the heavy modern jets and big enough to do so without the problems we had in Victorious. However, it seemed to me that if we could afford only one such ship it would suffer the problem of being too precious to risk and therefore something of a liability. By now I knew enough to keep views like this to myself and a few trusted friends but I need not have worried because CVA-01 was about to be sunk – by the RAF.

The RAF was a very powerful organisation politically and ran an effective public relations campaign that maintained a high profile for the RAF in the eyes of the public and traded very cleverly on the Battle of Britain and the fame of 'the Few'. The many thousands of aircrew who died in the bombing campaigns of the discredited Trenchard doctrine did not feature in this and the result was much public support for the RAF. Despite the evidence of the Second World War, the Korean war, Suez and other conflicts, the RAF adhered to the doctrine of independent air power and in furtherance of it was intent on eliminating the Fleet Air Arm. In this aim it had a great deal of political support, particularly from the Left.

The Navy, on the other hand, maintained its tradition of the 'silent service' and so played into the RAF's hands. The publicity-seeking activities of the RAF were despised by the Navy who felt that they had a good case and could stand on their proven record of successfully providing virtually all the air support for British forces in the various engagements since the Second World War.

The awakening came with the 1964 Defence Review, which featured a well-prepared attack on the Royal Navy and, in particular, on the Fleet Air Arm. The Navy was, in effect, to be reduced to little more than a coastal force. The Navy was invited to respond to the attack by defending itself in respect of a worst case scenario involving a war in the Far East requiring a convoy to be fought through to Australia.

At first it seemed an ideal scenario for the Navy because the crucial factor was the provision of air cover and we knew that the RAF could not provide it from shore bases as their aircraft had insufficient range. To everyone's surprise the RAF's position paper showed that not only could they provide the cover but that they could do it at less cost in terms of money and manpower than the Navy. The paper was well prepared, obviously researched well in advance and, at first sight, very convincing. The Navy was caught completely unprepared and in a panic reaction most officers at the Admiralty were co-opted to prepare a rebuttal of the RAF's case. Some of it was not too difficult. The RAF had been very selective in their data and much of it had been 'massaged' to an almost unrecognisable degree. There were no blatant lies but many unfavourable truths had been omitted. We could shoot it all down but the 'fact' remained that air cover appeared to be available throughout from shore bases. We knew that could not be true, so how had they finessed it? Eventually a sharp-eyed member of our team spotted it. On the charts the RAF had shifted Australia bodily 800 miles to the west to fit the range of their aircraft.

With this skulduggery exposed their case collapsed. But it made no difference; the government was intent on ignoring inconvenient facts. Within a few years the carriers were to go, the Navy's Phantoms and Buccaneers were to be handed over to the RAF and the Fleet Air Arm would consist of a few helicopters. And that is what happened.

17

Pause for Thought

With the carriers and their aircraft being run down it was quite obvious that a good many of us would have to go with them. Already the declining Navy was overburdened with senior officers and now there was a potentially enormous glut of talent in the middle ranks. Promotion, already difficult, was going to become a great deal worse. In those days you couldn't get out of the Navy unless they wanted to be rid of you. Now we were being offered financial incentives to go but the terms were less than generous. You couldn't choose your time, the Navy could refuse or delay your application and the fact of applying automatically made you ineligible for promotion. It was a bit of a gamble and a difficult decision. I foolishly thought I had a chance of promotion so I hung on.

My next appointment was to the Royal Naval Air Station at Yeovilton as the Senior Mechanical Engineer. There the first Phantom squadron had formed and was working up to seagoing capability, despite the sentence hanging over them. There was always the hope that something would happen to cause an outbreak of sense which would save the Fleet Air Arm from being handed over to the RAF in two or three years' time.

Meanwhile the RAF had taken delivery of the first Harriers, not with any great enthusiasm, and what for them was a bit of a sideshow.

With the coming of the Harriers and the scrapping of the carriers my interest in V/STOL concepts, which had, of necessity, been on the back burner, was now rekindled. Surely now, I thought, the Navy would be bound to consider operation of aircraft from small ships in a more favourable light. In fact I thought that there were signs that this was beginning to happen because the Navy had produced a requirement for an 'anti-submarine cruiser' that would operate helicopters. Preliminary drawings showed a ship that looked rather like a small aircraft carrier but with a flight deck stopping short of the bows. It was a bigger ship than I had had in mind but it was a step in the right direction. I felt encouraged by this sign to give some more thought to the possibilities of V/STOL.

At Yeovilton I lived in the wardroom and only got home at weekends when not on duty. I was often at a loose end in the evenings so instead of propping up the wardroom bar I began to investigate the possibility of launching a Harrier from a frigate. With hindsight this was an over-ambitious target but I reasoned that if frigate operation could be achieved everything else would be easy.

I knew nothing about the Harrier and its performance beyond what had been published – which I knew from experience would err on the side of optimism. The security regulations and the 'need to know' rule would prevent me from finding out what I needed to know about its true capabilities but I could guess that its payload and range performance from a vertical take-off would be very limited by comparison with a conventional aircraft or its own performance from a short take-off (STO). To be of any use to the Navy it would need, at the very least, its STO performance. How could you do this from a frigate? Obviously you would need some sort of catapult. But first I had to decide in my own mind what assumptions I was going to use about the aeroplane's performance. I had plenty of data about the Hawker Hunter, which the Navy used for training. That looked to be comparable in size and weight to the Harrier so I took that as a basis on which to build a hypothetical Harrier. To the basic Hunter I added a paper engine whose thrust could be whatever I wanted and to this engine I added the ability to deflect its thrust from horizontal to just past vertically downwards. Thus equipped I turned my attention to the catapult.

There was no question of using a steam catapult. Apart from the fact that it would be too big and heavy there would be no steam available in a gas turbine frigate. The catapult would have to be either a derivative of the old hydraulic catapult that had been superseded because of its inherent limitations or an entirely new type specially designed for the purpose. I took it upon myself to design such a catapult, which, again with hindsight, was very foolish. I just couldn't resist the challenge.

A vertical catapult looked attractive at a superficial glance as a space saver but it couldn't survive a closer examination. The aeroplane certainly would not like it and a few sums indicated that a surprisingly large amount of launch energy would be necessary to avoid a distressingly rapid return to earth. In any case the aeroplane had to end up flying horizontally so why not give it a bit of horizontal velocity from the start – in other words launch it at an angle away from the vertical. More sums, assuming a variety of different angles, yielded some interesting and quite remarkable results. As the angle moved away from the vertical the energy and speed required for a safe launch became less, not just a bit less but a lot less. From the horizontal the effect was even more marked. The first 20–30 degrees from horizontal gave really dramatic reductions in launch speed required. I thought probably not many people knew that so I would keep it to myself for the time being.

My discovery of the launch energy savings achieved by launching at an angle above horizontal eased the launch problem to such an extent that it would have been possible to return to the use of the old type hydraulic catapult but I decided to pursue the more interesting course of designing an entirely new type of catapult. I should have known better. The design I came up with was radical and risky but I made a working model that gave me confidence that, although it would need a great deal of development, might just be feasible.

At this point I suffered a nasty personal blow. I had seen it coming but that made its eventual arrival no easier to bear. In those days officers were promoted only in certain seniority 'zones'. Once you had passed the upper level of seniority you could not be promoted, however good your performance. That is what now happened to me; the gamble I had taken on getting promoted had not come off. In order to avoid the possibility of a passed-over officer causing embarrassment to authority by turning in an outstanding performance such officers were normally given undemanding dead-end jobs. So there I was aged 39, unable to get out of the Navy before age 45, and therefore faced with six years of static boredom. My morale was at rock bottom. Looking back on it from a perspective of 40 years it seems a silly reaction but it did not feel like that at the time. Ambition distorts one's judgement and makes a coward of you. Being passed over was actually a blessing in disguise. Well disguised.

In the middle of 1969 I was appointed back to the Naval Aircraft Department of the Ministry of Defence, which was now in Golden Cross House, opposite Charing Cross station in central London. The job was just the kind of dead-end job that I had feared and involved the control and approval of modifications to naval aircraft. On the positive side I could do the job with no great effort and with time to spare. I could use the spare time to get on with important things, in my view, like V/STOL and catapults. I also carefully set about trying to get some support for my ideas on catapulting Harriers from small ships and for my entirely new catapult design. My immediate superiors were supportive, up to a point, though not sufficiently for me. In retrospect they were very forbearing because they had their own careers to consider and I must have been quite a nuisance to them. Anyway they passed on my ideas to the Royal Aircraft Establishment at Bedford, which dealt with naval matters, and they arranged for me to see the chief designer of the Harrier.

My meeting with John Fozard, the chief designer, was a very instructive one. John was a larger than life Yorkshireman, which meant that he was a lot larger than life. I once heard him describe Yorkshire, to a visiting American, as 'the Texas of England'. He also had a first-class brain and a vast background of aircraft design knowledge. His blind spot was the operation of aeroplanes from ships, about which he knew little. He regarded shipboard operation as of little consequence for the Harrier. His reaction to my catapult ideas was expressed politely but in unmistakeably clear terms – John's views were always unmistakeably clear. He wanted no part of it and the Harrier had no need of it. The structural alterations needed to make the Harrier suitable for launching from a catapult would compromise its performance and cost millions. I was proposing unnecessary complications to an essentially simple aeroplane. He wanted to keep it simple. It is easy to understand and to sympathise with his reservations. The secret of the Harrier's success is the way that the designer has kept his objective firmly in sight and not been deflected by the temptations of the 'wouldn't it be nice' merchants who have so often compromised the designs of military aircraft. I know well that simplicity is not easy to achieve and not always appreciated when it is.

John's verdict was that the Harrier could do all that was required of it on shore or on a ship and it needed no external help. My catapult was an interesting and ingenious concept and he wished it every success away from the Harrier. Kindly he ushered me out but on the way he introduced me to Trevor Jordan and told him to provide me with some basic Harrier data. This was to prove to be a crucial introduction.

The catapult experts at Bedford also reacted and they were neither polite nor kind; at best they were condescending. They treated me as an ignoramus who knew nothing about catapults and ridiculed the concept. To be fair there was much to be critical about. There was a great deal of untried (at least in this context) technology and much use of new, and as yet unproven materials. It would certainly require a great deal of development effort. All this I was aware of but I had hoped that my input might trigger some constructive thought on the subject. What was really insufferable was their *de haut en bas* attitude. I've been patronised by better people.

The most deadly ridicule arose spontaneously and I have to admit that I walked into it in all innocence. I had produced a simple diagram to illustrate the working principle of my new catapult (Figure 10). My objectives, bearing in mind that I was thinking of

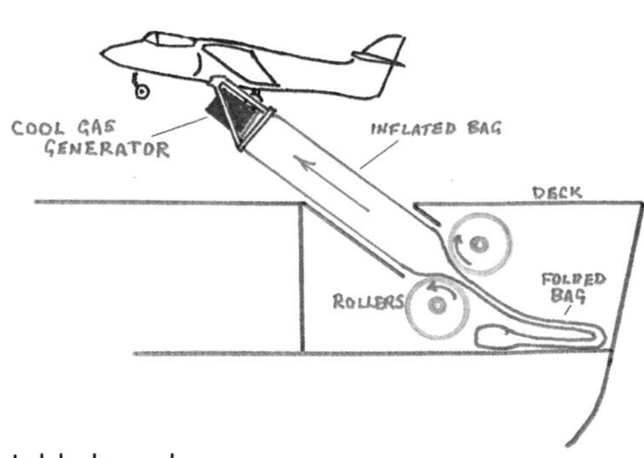

Figure 10 The inflatable launcher

a frigate installation, were to minimise space and weight. I was also aware from my own experience that one of the biggest headaches of catapult design was how to stop the moving parts of the catapult at the end of the launch. It helps if the moving parts are kept light. My new catapult pushed the aeroplane from behind and below instead of towing from the front as before. The motive power was provided by an inflated sausage-like bag running between rollers rather like a mangle. Inflating this bag fast enough at the required pressure was not going to be a trivial problem and preventing it from bursting under hoop stress meant careful choice of materials but none of these things seemed to me to be insuperable problems. Just a little bit difficult – a challenge.

So I had my blind spots, too, and I offer no excuse. To me it was simply a catapult with problems to be tackled; others saw different resemblances and soon it was known as 'the giant condom'. At least it created widespread amusement – it also killed the whole idea stone dead.

Actually, it went rather beyond that because I had stuck my head up above the parapet and attracted the notice of senior people in the operational hierarchy of the navy. They didn't like what they saw. Here was this mere 'plumber' daring to interest himself uninvited in what were essentially operational matters and, even worse, expressing his views on the subject. The Director of Naval Air Warfare (DNAW) himself intervened personally. He sent word that I was to shut up and get on with my proper job; above all I was never again under any circumstances to mention catapults in connection with launching any kind of aeroplane from ships.

The reason for this was the plans for the new anti-submarine cruisers, which were now given the title of 'through-deck cruisers'. The Navy was afraid, with good reason, that talk of catapults in that connection would provoke the RAF to oppose them as they had just successfully opposed the big carriers. So I retired hurt and angry at what seemed to me to be petty and obstructive politics that should not be tolerated in something as important as the defence of our country.

18

The Birth of the Ski Jump – 1969

There didn't seem to be much point in carrying on with my hobby and for some time I really did try to put it all behind me but the naughty thoughts kept nagging at me and wouldn't go away. Anyway, they were a lot more interesting than my proper job and anger is a great motivator. Not that I was going to waste energy on empty gestures and clearly catapults were right out of the question, for the time being anyway. So I would have to take a different approach, one that had crossed my mind several times but which I hadn't pursued because of my target of launching from a frigate. The inclined launch reduced the required launch end speed to such an extent that it was tempting to consider whether the aeroplane could do well enough under its own power alone to make it worth while. An uphill start was out of the question; there would have to be a horizontal run followed by a curved entry to the launch angle. In turn that meant that I should forget angles of 40 degrees or 50 degrees from the vertical, which would be quite impracticable. I would now have to think in terms of angles from the horizontal with a practical maximum of 30–35 degrees. The big question was would this give any worthwhile advantages compared with a short take-off?

My launching system now comprised a level run and a curved bit (Figure 11) and it couldn't be simpler in terms of hardware. Calculating how well it would perform was, however, anything but simple to one of my limited mathematical ability, which had already had a mauling at the hands of the boffins. I wasn't going to expose myself to that again if it could be avoided. I began to wonder if it was worth going on.

I really didn't have much choice about it, however, because the problem wouldn't leave me alone and kept nagging at me. As a result during my Christmas leave in 1969 I started making some calculations using the kind of straightforward maths that I could cope with.

THE SKI JUMP

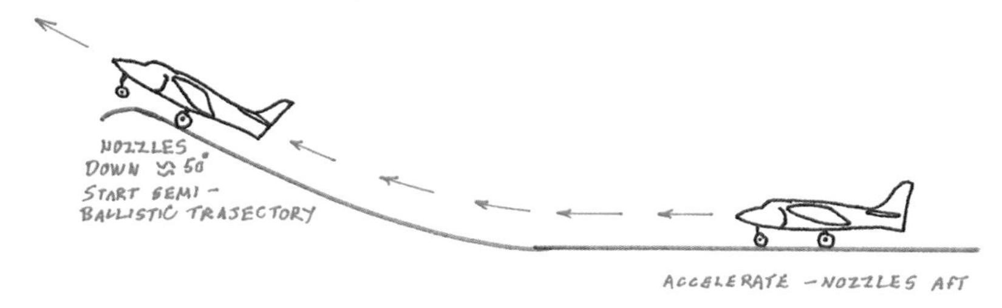

Figure 11 The Ski Jump in principle

I will go through these methods in some tedious detail because some boffins have suggested that the problem is trivial. I hope to show that it isn't.

First of all we have to calculate what speed the aeroplane will get up to along a given length of level run. I decided to do this by assuming a constant acceleration based on the aeroplane's thrust to weight ratio. This is not strictly the case but it is a fair approximation.

Now we have arrived at the curved bit and we have to decide what is the smallest radius we can use bearing in mind the speed of the aeroplane at this point. The limiting factor is the strength of the undercarriage and how much centrifugal force it can tolerate on the curve. I used figures provided by Trevor Jordan for this. We also need to calculate how much speed the aeroplane will gain or lose on the curved ramp. For this I assumed an average ramp angle – another approximation but good enough for present purposes. We add this to the speed we got on the level run, which gives us the ramp exit speed. Before we leave the ramp behind we must consider the fact that in travelling round the curve the aeroplane has acquired a lot of nose up angular momentum that may or may not be an embarrassment. Fortunately, when the nose wheel leaves the end of the ramp the aeroplane is subjected to a very powerful nose down acceleration, which, in the interval before the main wheels also leave, is sufficient to cancel the nose up motion. In fact subsequent calculations for aircraft other than the Harrier show that this seems to be generally so and all aircraft are left with little or no residual pitching motion at this point.

So now the thrust is deflected downwards through an angle of 40–50 degrees and the aeroplane has begun its semi-ballistic launch trajectory at an angle above horizontal. From now on everything is changing except the thrust and the weight of the aeroplane. The aircraft attitude and the angle of attack will change continuously as will the airspeed and the flight path. The way I chose to tackle this was to resolve the forces acting on the aeroplane – thrust, weight, lift and drag – into vertical and horizontal components from which accelerations could be calculated. These, in turn, yield changes in velocity to be added to the previous conditions. This process is then repeated for conditions a short time later and then again and again at the same time interval until it is found that the aeroplane is beginning to climb away under a combination of wing lift and the vertical component of the deflected thrust. I called this the 'fly away point' as it marks the end of the take-off process.

This was before electronic calculators and computers were available to one and all. The tool I used was a slide rule. For those lacking this dubious experience the slide rule was a kind of mechanical calculator capable of multiplication, division and working out trigonometric functions. It couldn't add or subtract or tell you where the decimal point should be. It was all too easy to lose track of the decimal point and thus be in error by orders of magnitude. Using a slide rule demanded good eyesight and was incredibly laborious. It would take me a week of spare-time work to do the calculations for just one launch – and that is if I didn't make too many mistakes. Even then, in the early stages, I might find at the end of the calculation that the launch was a disaster ending in a splash because I had chosen too short a run or too heavy a weight, or the opposite extremes. Very soon, however, I began to get a feel for things and made a very encouraging discovery: take-offs using this method, with no catapult assistance, were even better than I had dared to hope and they needed much less space than a STO.

The launch sequence is shown in Figure 12 and compared with the then conventional STO that the Harrier used. You couldn't use this method to launch from a frigate but you could use it conceivably to launch from a *frigate-sized* ship.

This was over Christmas 1969 and it was all good stuff. In fact it was too good for me and left me deeply suspicious because I'm an engineer and I know that nothing is for nothing; there must be a downside, a penalty and so far I hadn't found it. The natural conclusion, given my shaky mathematics, was that I had made some fundamental error either in my calculations or my assumptions. Needless to say I kept all this to myself. I was developing a thicker skin but there was no point in courting ridicule.

The weeks went by and the suspicion nagged at me. I checked my assumptions and repeated my tedious, boring calculations and still things came out much the same. My simple, catapult-free launching method was much, much better than an STO but I couldn't convince myself of that because it appeared to be something for nothing. Meanwhile I had my 'proper job' of sorts, which occasionally claimed my attention.

In early 1970, with no particular effort on my part, my elusive penalty revealed itself or, to be more truthful, I noticed it because it had always been right in front of me. The penalty was that my launching method took a considerably longer distance from brakes off to the fly away point than the STO, but of course very little of this distance was on the ground. I called it the 'Runway in the Sky', which I thought summed up the principle admirably.

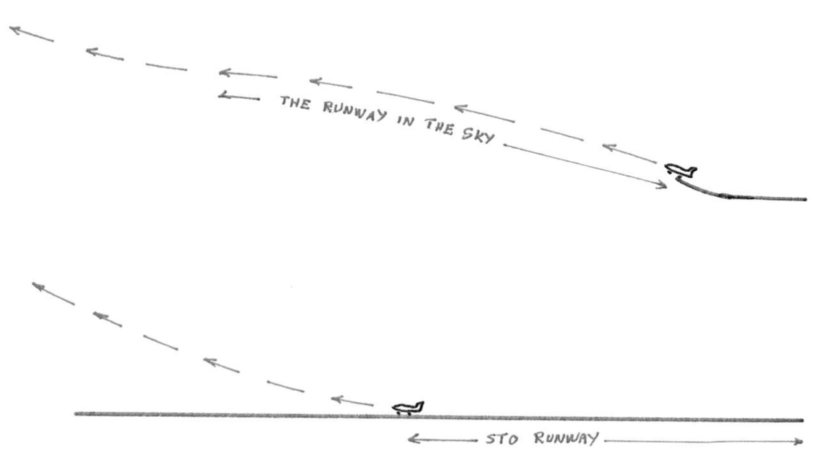

Figure 12 The Runway in the Sky

The relief from the nagging doubts was enormous and it seemed that the last hurdle had been cleared. I had a system that would do most of what I had set out to do, which overcame all the objections and which should be fairly straightforward to prove. This was probably the time to reveal it to the Navy. This time they might be delighted.

19

Trying it On, 1970–1

By this time I had taken some of my friends into my confidence and also discussed it, very casually, with my immediate superiors. They were all engineers and were prepared to listen to and understand my reasoning. They were generally supportive, up to a point, but I suspect that they all wondered why I was persisting with the idea and some told me frankly that they thought I was wasting my time. With hindsight I realise that they were being very tolerant – they had their own careers to think of. However, I thought it was worth trying it on the operational people and DNAW because, after all, no catapults were involved. They were, in fact, a good deal less than delighted. I doubt if anyone on the operational side even bothered to read my paper on the subject; certainly the response that came back showed no sign of understanding of my arguments. It was back to the old refrain that I was putting the through-deck cruisers at risk and anyway the Sea Harrier would need no help.

Yes, the Sea Harrier. The Navy had succeeded in getting approval for its own version of the Harrier to go with the new cruisers so it seemed to me that the RAF had no great objection to that turn of events. I do not know why, after battling so hard against the aircraft carriers, the RAF had apparently relented but I suspect that it was because they didn't consider the Harrier to be a 'proper aeroplane' – it was, after all, subsonic and it landed vertically. Real men flew supersonically and screeched in to land at 120 knots or more. It has to be admitted that many, if not all, Naval aviators shared that view. The Sea Harrier was a second best, acceptable only because they were denied the big carriers and big supersonic aircraft that they really wanted. And now here was an engineer of all people saying that they could use ships even smaller than the cruisers. They were decidedly not delighted; the answer was still no, only more so.

With no support from the operators, my Runway in the Sky was going to fade away into oblivion unless it got backing from the boffins at Farnborough and Bedford and in view of my previous experience with the Giant Condom I was very reluctant to approach them again. My boss persuaded me to have another go because, after all, it was on the face of things an entirely new concept and it would be churlish of me not to take advantage of his support. Once again a paper was prepared and submitted with the same negative result. They were particularly scathing about my calculations, the methods I had used and my assumptions. What was especially objectionable was the statement that an assessment of the concept was not worthwhile because it had 'negligible intellectual content'.

There can be no denying that my mathematical methods were amateur and that I had made assumptions that would lead to inaccuracies but Bedford had the expertise and the wherewithal to check my work by their own approved methods. Had they done so they would have found that my inaccuracies worked against me and my claims were, if anything, understated. Above all, they had no right to dismiss out of hand the work of a professional in that particular field. However, they had done me a favour because I was very angry at the insults and in the years to come, whenever frustration and flagging energy made me think of forgetting it all, I had only to recall the anger to find the energy to continue.

I was back at square one but this time with growing support from colleagues and superiors who shared my resentment of the cavalier treatment I had received. In particular Admiral Dyer-Smith, the head of the Department, decided that more effective action was required. He said, 'We have got to make you academically respectable'. I had no idea how this could be done and, in fact, suspected that it probably couldn't be done. My confidence in any academic ability that I might once have had was at a very low level.

The Admiral went ahead by proposing that I should have a year's sabbatical at a university to do more work on my idea under academic supervision. He persuaded Professor Ian Cheeseman of the Department of Aeronautics at Southampton University to accept me. The Navy was rather dubious about the idea and the Director of Naval Educational Services, invited to comment, considered that I was 'unlikely to make any significant contribution to the Naval service'. Nevertheless the application went ahead.

Meanwhile the wheels of the naval appointments system ground on and found another job for me as the First Lieutenant of the RN Barracks at Chatham. This was a typical dead-end job for a passed-over officer but not an uncongenial one. It didn't stretch the grey matter but it had one great merit in my eyes, which was that it included running the Sailing Centre in the dockyard. I enjoy sailing and it caused me no pain to spend many afternoons messing about in boats – and mornings, too, from time to time. Under my tender administration the Sailing Centre blossomed, became smarter, better equipped and well used. It impressed senior officers more than anything else that I had done in my career and all with comparatively little effort on my part. Of course it could do nothing to advance a career that was, in effect, over.

In fact I would be allowed to leave the Navy when I reached the age of 45 in two years' time and at this stage the Captain of the Barracks told me that he would like me to remain there as First Lieutenant until my time expired and that he could arrange that, if I would like to stay. I could have a very agreeable final two years and it was a tempting prospect. Only a day or two later the MOD responded at last to my application for a sabbatical year at Southampton University and rather to my surprise, in view of the comments of the Director of Naval Educational Services, agreed to it. There was, of course, a snag. In exchange for that year the Navy would require me to serve for a further five years to age 50.

That really put me on the spot. I reckoned that it was a hard and unreasonable bargain, sanctimonious and grudging in the circumstances. Delaying my release from the Navy by seven years would reduce my prospects for civilian employment and there was no guarantee that my achievement of 'academic respectability' would result in any more success than I had had in the past. On top of all that, the scathing comments of the boffins and the confident prediction of DNEdS that I was unlikely to do the Navy any good had done nothing for my confidence. Perhaps what little academic ability I might have possessed really had trickled away over the years. I was sure that I was right about the benefits of my 'Runway in the Sky' but not at all sure that I would ever be able to convince the operational people of its merits. As I had never been to a university that also was a bit daunting. The image I had in mind was formed almost entirely by the works of Evelyn Waugh.

A pleasant two years at Chatham and then freedom had great appeal but there would be one certain result – the 'Runway in the Sky' would never happen. There really wasn't any choice. I had to back my own idea and Admiral Dyer-Smith's confidence in me and so I went to Southampton.

I was very glad that I had done so because a new world opened up to me. Whatever my colleagues thought of my academic qualities they forbore from comment and accepted me at face value. They even seemed to consider that I might have some ideas worth listening to. Emboldened by this unaccustomed tolerance I became quite reckless in floating some thoughts around without provoking howls of outrage. Possibly that was because they had no idea of the peculiarities of life with aeroplanes at sea. I had, of course, done all the basic work on my celestial runway; now I had to translate that into academically respectable terms. But first I reckoned that I needed a shorter, snappier title for the idea because 'Runway in the Sky', while totally appropriate, was rather cumbersome and hadn't caught on. So I called it the Ski Jump because of the obvious similarity.

One of the first things I had to do was to learn to use the university's computers. These were very primitive by today's standards and not at all user-friendly. Since I would need to write my own programs I enrolled in a fortnight's course in Fortran and emerged in a baffled but willing condition to develop equations of motion.

After a great deal of brain-bursting effort I brought forth, not without considerable misgivings, four simultaneous differential equations. I hate differential equations but there they were and I had no one but myself to blame. The problem was, how was I to solve them? I would need advice for that but that in turn brought another problem. Would the sight of my pretentious efforts reduce any advisor to helpless laughter? There was only one way to find out so I tried them on several people. The general consensus seemed to be that they looked OK and the next move should be to try them on The Computer. This lived in the Mathematics Department and was about the size of a large garden shed. One's problem was fed into The Computer by means of punched cards. My equations amounted to an impressive pile of cards, although I found out later that it didn't take much program to make a pile. During the day The Computer was reserved for members of the Mathematics Department but we lesser beings were allowed to feed it with our puny offerings after five o'clock, when it would digest them overnight and disgorge the results in the morning. That was the theory. In my case I invariably got my stack of cards back with a slip inserted, usually near the beginning, saying 'error'. It didn't say what the error was or where it was. To this day my own computer does this to me occasionally and I find it particularly unhelpful and infuriating. After about a week of unrewarding toil I took my problem to one of my mentors. 'I've got just the thing for that,' he said. It was called an analogue computer (I don't think they have them now) and it looked like a sheet of peg board into which you could plug various devices labelled 'integrator', 'differentiator' and so on. Everyone should have one. When we had it all set up the back of it was a most impressive cat's cradle of wiring. All that remained was to connect it up to a cathode ray tube, switch it on and it should display my Ski Jump trajectories for various weights and launch conditions. And it did! Vindication! Armed with this I could range over the whole gamut of likely conditions in no time at all but there was a snag, as usual: the analogue computer was booked up for the foreseeable future.

This was a severe blow but my friend was full of good ideas. First he had another look at my equations and muttered words like 'highly non-linear'. He opined that the equations were 'not amenable to an analytic solution' and advised me to go for 'arithmetic treatment'. I hadn't the foggiest idea what he was talking about. He said it was no problem – 'Give it the old Runge-Kutta procedure.'

I confessed to my ignorance and he explained it to me in great detail. The trouble with people of a mathematical inclination is that they are quite unable to understand the mental processes of normal people and they freely use obscure jargon. Apart from that I just wasn't clever enough to take it in. Very politely, I asked him to run through it again and, very patiently, he did. I still understood none of it but I hadn't the nerve to ask for more. 'Thanks a lot,' I said. By way of parting advice he suggested that I use Basic to write programs because it would highlight mistakes as I made them and was almost the same as Fortran.

I went away very worried. Messrs Runge and Kutta had defeated me. The analogue computer had confirmed that I was on the right track but there was an awful lot of work to be gone through before I had covered sufficient conditions to make a convincing case for the Ski Jump. And I couldn't keep on bothering busy people with my queries. On the other hand, here I was and I had got to get on with it somehow. At least I knew that the Ski Jump worked as advertised. It was a matter of calculating the numbers and my original work had given me a good idea of their approximate values. I decided to abandon the differential equations and revert to my original laborious method using a slide rule but this time The Computer could do the work. Instead of taking just three or four points in the trajectory The Computer could do it every second or even more frequently if I felt so inclined.

So, using Basic this time, on punched paper tape, I wrote a program that resolved the forces acting on the aeroplane and turned them into accelerations and then velocities and then displacements at very short intervals throughout the trajectory. After the usual business of finding and correcting my non-deliberate errors, it worked. The results were consistent with the few I had done on the analogue computer and now I was confident I had a tool that I could use for all the rest of my work. Crude it may be to the purist but my method was good enough for me. Busting with pride in my great achievement I showed it to my mentor, rather expecting him to be a bit snooty about it. 'Ah, yes,' he said. 'I told you so, the good old Runge-Kutta technique.'

From this point on I was able to progress quite rapidly in preparing a respectable case for the Ski Jump covering all the likely launch conditions. The results were, as I had come to expect, very favourable. Typically, in representative conditions, the take-off distance was reduced by about 70% compared with a normal STO. If the Ski Jump was combined with a low-energy catapult then the space required to get airborne at STO weights was very little more than that required for a VTO. I reckoned that I now had a good enough argument to demolish any remaining opposition. However, I was well ahead of time with quite a lot of my sabbatical year left so I decided to make it even more complete by examining various possible alternatives to the Ski Jump by way of comparison.

Having done this I was confident that the technical case was beyond any logical argument but I was still unhappy about the Naval operational people. I anticipated (correctly) that no one of any consequence would bother to read my technical arguments so I should try to appeal to them on operational grounds. I assumed that in the light of the recent successful attack by the RAF they would have been forced to accept that big carriers were no longer an option and that operation of aircraft from small ships was a subject worthy of special attention. Accordingly I added a section to this effect to my thesis. That was a mistake and I should have known better.

However, the university was happy enough with my thesis and awarded me the degree of Master of Philosophy, which was a very pleasant surprise to one of my unacademic background. There was one slight impediment that I thought was depressingly typical of our times. I had originally given my thesis the title 'The Operation of V/STOL Aircraft from Small Warships' but my supervisor advised me to amend the title to '...from Confined Spaces' because there was a strong pacifist element in the university hierarchy who would otherwise put obstructions in its way. All in all I enjoyed my year at Southampton and will always remember with gratitude the help and support I received from colleagues and staff, especially Professor Ian Cheeseman. Now it was back to the task of convincing the Navy that I had something very well worthwhile to offer.

20

Trying it On Again, 1973–6

I went back to the Aircraft Department in Golden Cross House, but this time to a rather more active and productive desk in Future Projects. One of our main tasks was the development of a new helicopter that had the imaginative title of Sea King Replacement. This started life as a paper study for a twin-engined helicopter to do the same jobs as the Sea King but better. As such there should have been no great difficulty in formulating the details of the requirement and the hardware to bring it to fruition. The trouble was that everyone could think of more and better things for it to do and more and better equipment with which to do them. The result was that the helicopter became heavier and heavier while still at the paper stage and as a consequence of this obesity it needed more power to get it into the air. Eventually, when the weight hit 20 tons and the two engines became four, common sense intervened and the papers were torn up. Some of us thought that the Sea King Replacement would have to be a Sea King. But with a clean sheet of paper we would start again, only this time we would be ruthless in resisting the siren calls to add more goodies to the aeroplane. Despite our efforts the process was repeated over and over during my time in Future Projects and the helicopter remained anchored to paper by its inexorable addiction to 'nice to have' and weighty additions. Long after I left the Navy it actually appeared as a real flying helicopter called, appropriately, Merlin as it needed a magician to get it to that stage.

But that was all in the future. My objective on joining the Department was to get something moving on my own project. With this in mind I spent much of my first few weeks distributing copies of my thesis to anyone whom I thought might usefully read it and in preparing a presentation of my findings on the performance to be expected from the Ski Jump. In February 1974 my presentation took place in the Old Admiralty Building before an audience that included DNAW and other operational staff, plus representatives of Farnborough, Bedford and British Aerospace. I had high hopes of it.

Reactions were generally not as negative as they had previously been but they were not particularly encouraging from my point of view. DNAW and the other operators were cool. It was an interesting concept, they thought, and I was to be commended for my efforts in the direction of technical advancement but, in fact, they were not needed and all the pernicious nonsense about operating aircraft from small ships was unhelpful to the Navy at a time when the Sea Harrier had not yet received final government approval. In effect a patronising pat on the head coupled with a kick in the backside.

So far as the Farnborough and Bedford boffins were concerned they were prepared to concede a limited degree of academic respectability and there were no more scathing comments. What we now had was long range sniping. For example, 'Not all the projected benefits may be realisable,' and 'What happens if there is an engine failure during launch?' British Aerospace were much more positive but at the same time very cautious. In no circumstances were they going to suggest that the Sea Harrier would need any form of take-off assistance except, possibly, in the circumstances that I had postulated and the Naval Staff had firmly rejected. However, I believe that they saw the possibility of a nice little study contract that might turn out to be useful at some time. They were right about that and in due course a study contract was arranged to investigate the possible benefits of the Ski Jump.

For the time being at least there wasn't much more that I could do. Things had advanced a little bit if not as much as I had hoped and I would have to be content with what had been achieved. At least now I had strong support from fellow engineers within the Department, notably from Commanders Bowles, Ashmead and Jenne who, in the months to come, put time and effort into helping things along. There were times when their encouragement was especially valuable.

The study contract was, by Ministry of Defence standards, a minor one and, though welcome, was not a matter of urgency for British Aerospace. It certainly didn't move fast enough for me;

in fact at times it seemed to me that it wasn't moving at all. My contacts at the British Aerospace works at Kingston were Trevor Jordan, in charge of Harrier performance estimation, and Ken Causer, his deputy. They were making what I reckoned was an unnecessary fuss about the strength of the undercarriage. My sums said that the undercarriage was more than adequate to stand up to any loads that the Ski Jump might impose upon it and there were other, more important, aspects to be checked. So Doug Thorby, the undercarriage specialist, was called in and an unwelcome truth was revealed. The figures I had been given were incorrect and the undercarriage strength was a lot less than I had been led to believe. We bullied poor Doug Thorby but after conceding a little he stood his ground and gave us some limiting figures that we could use. The result of that was that, unless the undercarriage was strengthened the Ski Jump ramp radius would have to be increased to ease the applied loads. This, in turn meant that launch angles of 30 degrees or more would not be practicable because the ramp would become too large to use on a ship and even more so on a small ship. In fact even 25 degrees might be stretching things a bit too far. However, there is a law of diminishing returns that applies to the launch angle (see Figure 13) so even if we were limited to less than 20 degrees the performance loss would not be disastrous. We would still be getting most of the possible gain.

With these restrictions in mind, Trevor and Ken set about developing computer software to do the performance calculations we wanted, which would take into account factors of which I had so far been blissfully unaware. Such things as lift loss due to the air flow induced by the downward deflected jet thrust – one of those malicious little jokes that nature is all too ready to perpetrate. Another unanticipated (by me) effect was air intake momentum drag, which tends to destabilise the aeroplane – a quite unnecessary complication but amenable to treatment once you know about it. The undercarriage strength was another such but the Ministry of Defence was too parsimonious to spend the relatively small amount of money needed to strengthen the Sea Harrier's undercarriage and so it remained a limitation to the end of its days.

This study contract was, quite understandably, a bit of a sideline for British Aerospace. It would get done and there would

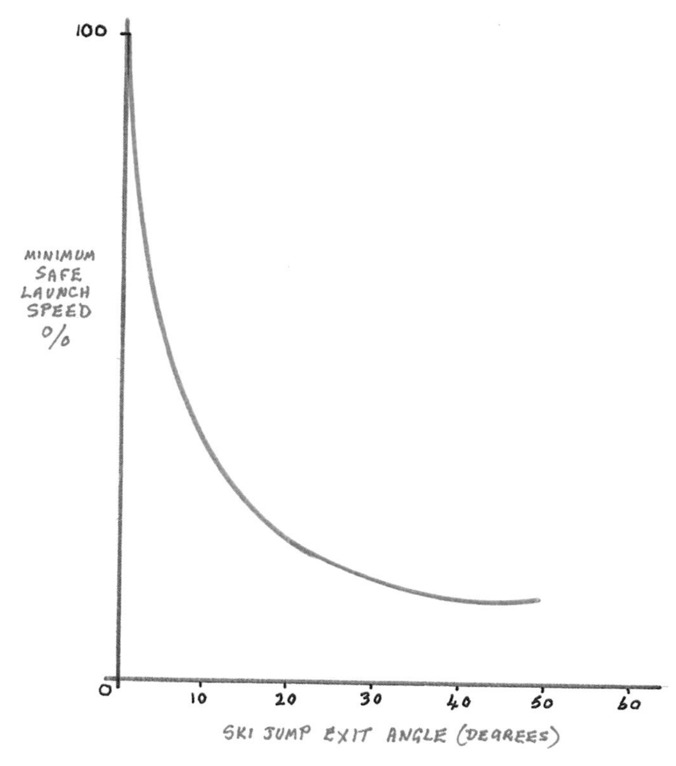

Figure 13 Launch speed plotted against launch angle

be a report in due course but I began to worry that the MOD would meanwhile do to the Ski Jump what it all too often did to inconvenient matters, that is to bury it quietly, over time, under a mountain of inertia and inactivity. I had by now worked in or with the MOD several times and had seen it happen to other projects. Rightly or wrongly I felt that I would have to do something to keep the project alive, but what? I decided that some external publicity was needed but here I came up against a problem that was largely of my own making but compounded by the Official Secrets Act. For Naval use, the Ski Jump was inextricably tied to the Sea Harrier and the performance data for the aeroplane were quite highly classified – confidential or above. If I wasn't very careful in what I had published I might find myself playing a central role in a court martial. I eventually found a way round the problem by what I fancied was a devilishly cunning plan.

I wrote an article entitled 'Payload Without Penalty', which showed how you could launch a hypothetical V/STOL airliner from a Ski Jump and get a vastly improved payload. No such V/STOL airliner existed then or now but the more knowledgeable readers might have noticed that the more important characteristics of my imaginary air liner, such as thrust/weight ratio, were uncannily similar to those of the Harrier. I sent the article off to the *Journal of the Royal Aeronautical Society* and, rather to my surprise, they accepted and published it. I have to admit that I was quite flattered by this acceptance and began to wonder if it meant that I had become a recognised boffin. In any event I thought that the article would generate discussion and controversy, which was what I wanted, so I sat back and waited for the uproar to begin. The ensuing silence was profound and prolonged and it brought home to me how difficult it is to beat determined indifference and apathy.

At about this time the Navy sent me off to NATO headquarters in Brussels to make my presentation on Ski Jump to an Information Exchange Group. I was not unaware of the motivation behind this; it was the semblance of action, not the substance, with the possibility in mind that the Americans might take an interest and relieve the MOD of any obligation to do something about it. I went along with it because any chance, however unlikely, was worth a try. Two officers from DNAW came with me, whether as 'minders' or not I've no idea, but they obviously thought that the whole thing was a bit of a joke. As it happened things turned out as I had expected. The Americans did show a limited amount of interest and their reaction was one of friendly encouragement together with some surprise that we were not going ahead.

It seemed now that after all the high hopes following my year at Southampton University and my subsequent presentation that I wasn't getting anywhere. I hadn't completely given up because I knew, with no false modesty, that the Ski Jump was potentially a concept of major importance in the context of naval aviation. The trouble was that no one else could see any need for it. I had failed to get the message across and I would have to put up with it. It was bearable and I had much to be grateful for.

21

Fortune Smiles, 1976

This is a good point at which to pay tribute to fortune or luck in all humility. We have no hesitation in blaming luck, bad luck, when things go wrong but we feel entirely justified in claiming all the credit for success. In this case I must acknowledge here my debt to fortune because I know well that without the good luck that now came to the rescue the Ski Jump concept would have died and never become reality, despite my efforts and those of my friends.

The first results of Trevor Jordan's and Ken Causer's work had just become available and they arrived in parallel with the discovery of some serious deficiencies in the predicted performance of the Sea Harrier when operating from the through-deck cruisers. The first of these was being built and her name was Invincible.

Fortune's intervention manifested itself in the form of a telephone call from John Fozard. The conversation went something like this.

'Doug, do you know what you've done?'

'What've I done now?'

'You've saved the Sea Harrier.' (Being a Yorkshireman, John was inclined to dramatise.)

Although the Sea Harrier had yet to fly it was possible to make accurate performance predictions by extrapolation from the existing RAF Harrier and these predictions showed that the Sea Harrier would not meet the specification requirements when making an STO from Invincible's deck. The shortfall was serious enough for there to be talk of cancellation. What was worse, if the ship was pitching the STO performance was so compromised that the Sea Harriers would have to make vertical take-offs with consequent large reductions in payload. Coincidentally, the results of the Ski Jump launch simulations, now rolling off the computer, showed that, using the Ski Jump, not only would all the specification requirements be met, they would be comfortably exceeded. And there was more: ship pitching would have little effect on Ski Jump launches.

Luck was exceptionally ladylike on this occasion because problem and solution were presented together.

The Chief Designer of the Harrier felt strongly that the aeroplane needed no external help to get airborne from a ship. Indeed it was one of the Harrier's unique selling features and one on which he was not inclined to compromise. But Trevor Jordan was one of the people whose views carried weight not only with John Fozard but also with John Farley, the chief test pilot. John Farley had read my thesis and was, to say the least, sceptical about it for reasons of which I was unaware. He had just returned from a visit to the US Marine Corps who had found that when doing STOs at heavy weights from a level flight deck the pilots were finding that from time to time the control column would come up against the forward stop. In other words, they were about to run out of forward control power. This was largely due to the upward airflow that occurs at the forward end of a carrier's flight deck. Now here was a person proposing that this aeroplane, with very limited nose down control power in these conditions, be required to run up a curved ramp that would impose a very rapid *nose up* pitch rate on it. He is a courteous man but not unnaturally his opinion of he who would suggest such a thing was terse and not in the least polite. John, in turn, was unaware that my calculations showed that the high pitch rate on the ramp was cancelled in the interval between the nose wheel leaving the ramp and the main wheels leaving because of the very high *nose down* pitching moment applied during this interval. In fact in all conditions the aeroplane leaves the Ski Jump with little or no residual pitching motion.

When Trevor pointed this out and showed that his simulations confirmed my predictions both Johns were convinced and for me things were never the same again because there now followed a truly impressive (to me at any rate) double act.

22

Trials and Triumph, 1976–8

The two Johns, Fozard and Farley, laid on a presentation in the Old Admiralty Building where I had given mine back in early 1974. The audience was as before but this time I was part of it. The ground covered was much the same as mine had been except that they, very wisely, made no suggestion that smaller ships could be used. They concentrated on the truly astonishing performance gains that the Sea Harrier would enjoy when operating from the Invincible-class ships. Professionally produced graphics helped of course but the combination of Fozard covering the more technical aspects in a decidedly flamboyant style plus Farley's low-key, almost understated, address of the pilots' interests was convincing beyond argument. The Harrier's chief designer and the chief test pilot had to be taken seriously and from this time onwards DNAW and his staff became strong supporters of the Ski Jump. Suddenly I felt like someone pushing hard at a door that is suddenly opened from the other side. As it turned out there was plenty of opposition still to come but bigger guns would be fighting it. I was, in effect, sidelined because the Navy gave me no official part in further Ski Jump development but I had no doubt that the project was in safe hands and I must record with gratitude that British Aerospace always gave me full credit for the original concept and the proof of its feasibility.

Following on from the joint presentation, British Aerospace was awarded another contract for further studies of the Ski Jump to decide what practical trials were required as a preliminary to making a final design for fitting to a ship. The Navy's designated experts on all things involving aeroplanes in combination with ships were, of course, the Naval Air Department at RAE Bedford who now fell into the role of supervisor for this contract. This was a natural and logical outcome, given the way the system worked, but it was not one that I could view with any enthusiasm at all in the light of recent events and their open hostility to the concept. There was nothing I could do about it and my worst fears were realised when a notable opponent of the Ski Jump was appointed as chairman of progress meetings. However, he and his kind now had to be more circumspect and the tactics they adopted were to be purveyors of gloom, casters of doubt and creators of delay. They could be quite cunning in formulating disaster scenarios to which there could be no convincing reassurance. For instance, what if a tyre bursts on the Ski Jump? The only possible answer at that stage was that we didn't know. This engenders an atmosphere of uncertainty. Actually, we now know the answer to that particular 'what if' because it happened. You get an unsightly black streak on the Ski Jump and a strong smell of burning rubber. That's all. I have to admit that if I had had any position of influence on the project I might well have intervened disastrously during this time because I became thoroughly fed up with the RAE and what I saw as obstructive behaviour. Luckily British Aerospace avoided confrontation and played a much more subtle game, which, while it didn't entirely avoid the obstructions, at least minimised their effects. For a long time the RAE insisted that there was no point in using an exit angle of more than two or, at most, three degrees for no good reason that I could see except that it would minimise the benefits. This was infuriating for me but British Aerospace gently shrugged it aside.

As I have said the Navy had not given me any part in this further development team. I don't think that it occurred to them to do so. We all had our places. Naval engineers were not boffins and the job was now in the hands of boffins and so, even though I was the originator, it was now nothing to do with me. So it was back to my 'proper' job of replacing the Sea King. Bedford were not obliged to invite me to progress meetings and they didn't but British Aerospace kept me informed and I just turned up. My immediate bosses turned a blind eye to my absences on these occasions.

You may wonder, as I did, why the project and sometimes I were subjected to this animosity. I had had no previous contact with any of the Bedford personnel until I came up with the Ski Jump idea and even then the contact was formal, by letter, not face-to-face. What could I possibly have done to cause offence? I found out very much later. It seems that Bedford had obtained approval and funding for a lucrative long-term study to design and develop a vertical catapult but then I came along and all unwittingly showed that this would be a waste of time and money. They were understandably not very happy about that but even so I felt that there was a disproportionate amount of bile that, even now, is difficult to justify. There was also of course the Not Invented Here (NIH) factor that predominates in all large organisations and thrives particularly in government research establishments. In this case I believe that the NIH factor was the prevailing emotion and as they saw it an interloper was trespassing on their territory.

All the same, things moved on and eventually approval was given to go ahead and build a trials Ski Jump at RAE Bedford and progress began to accelerate. We were not home and dry yet, however. The next problem was where was the money going to come from or 'Who's paying for this?' because, financially, the Ski Jump project did not exist. It had not originated from a naval staff requirement and hence didn't appear in the long-term costings and so simply wasn't in the system. This was, I was told, a major problem not to be taken lightly (I had politely asked how they coped with a war that wasn't in the long-term costings). To this day I don't know where the money was found. It wasn't very much anyway – less than £200,000, if I remember correctly – which, even then, was peanuts by MOD standards. The money was found and we could go ahead, except that there was another Bedford generated problem – they couldn't find a safe site for it.

This seemed to me to be all part of the RAE's surly and grudging acquiescence in the programme. They continually pressed for the launch angle to be very small (two to three degrees) and to cast doubt on whether the projected performance gains would be achieved, even though they produced no evidence to support their doubts and their claim that any gains would be marginal. Now there was this site problem. Eventually a site was found on a section of long disused taxiway subject to an inconvenient condition – the Ski Jump must fold flat so that vehicles and aircraft could go over it if necessary. (It never was necessary.) This meant that the Ski Jump had to be built over a pit as well as being variable in exit angle and profile. Despite all these petty hindrances things moved ahead pretty well until Bedford was joined by a powerful ally in opposition, the Director General Ships. He was the head of the warship design department in the MOD at Bath. I have often wondered what lay behind his hostility to the project but I believe that it was essentially a sense of solidarity with his fellow civil servants at RAE Bedford plus indignation at my and British Aerospace's trespassing on what he considered to be his territory. Whatever the reason, he now inserted his personal spanner in the works claiming that, practically speaking, the curved ramp could not be built to the required accuracy. This despite the fact that BAe were by now doing just that, adjustable and folding flat as well. In addition he added his voice to claims that any gains would be marginal (despite his total lack of expertise on the subject) and that the 'so-called Ski Jump' would not be fitted in any of the Invincible-class ships. I particularly resented the 'so-called' bit. To cap it all he issued an edict that all further investigation into Ski Jump performance gains was to cease (see Appendix 6).

This was a serious setback but fortunately it was very short in duration because, all of a sudden, we had very powerful friends – the Admiralty Board itself. The Controller of the Navy intervened personally to override DG Ships' edict on both counts. DG Ships was anything but magnanimous but now could be little more than a nuisance. The project continued fast and by mid-1977 the trials ramp was finished and the trials could begin.

They started in August 1977 at a six-degree exit angle. Before doing the first run up the ramp, John Farley considered that it would be prudent to do one or two runs alongside the ramp to see if the business of selecting nozzles down at the end of the ramp presented any difficulties in terms of timing. Accordingly he positioned the Harrier to one side of the ramp and accelerated as for a take-off. When abreast of the end of the ramp he selected nozzles down and the result was most spectacular. As the nozzles went down and the blast impinged on the surface, the taxiway appeared to erupt. Great sheets of tarmac leaped into the air and appeared to envelope the aeroplane but fortunately none was swallowed by the Harrier's engine. The surface of that long disused taxiway was obviously in a very dangerous condition and quite unusable with

Figures 14 and **15** Ski Jump trials at Bedford

deflected nozzles. Fortunately John decided that his one run had shown that there was no difficulty in timing the nozzle selection and so all further runs could be made up the Ski Jump, thus avoiding the tarmac problem. This was the last impediment and from this time on all went well. There were no nasty surprises and some very pleasant ones.

I had not been in the least nervous about the outcome of the trials. We had had to answer so many 'what ifs' from the various Jeremiahs who had inflicted themselves on us that we had covered just about every eventuality conceivable. The event was almost a bit of an anti-climax. Among the pleasant surprises was the revelation that there were no difficulties in controlling the aeroplane. John Farley described it as 'the easiest, safest and most high-performance way to get airborne that I have ever experienced in any type of aircraft'.

From the beginning I had always hoped that the opportunity would occur for me to join that privileged company who were among the first to experience a Ski Jump launch. In view of my unofficial status I broached the subject with some diffidence and I was lucky that when a two-seat Harrier appeared on the scene I was first in the queue.

Figure 16 Front, left to right: Doug Taylor, John Fozard, 1st Sea Lord (Photo: British Aerospace)

Figure 17 Left to right: Doug Taylor, John Fozard, 3rd Sea Lord, John Farley (Photo: British Aerospace)

Which brings us to the start of this story. I was about to take part in the test of my own invention. I had gone through it in my mind a good many times and often wondered how the fact would compare with my imagination.

'Off we go, then.'

The engine note rose as the throttle was slammed open, brakes off as the Harrier started to move and the grey steel wall of the Ski Jump seemed to hurl itself at us. One thing my imagination had not encompassed was the incredible horizontal acceleration of which the Harrier was capable. I had flown in a wide variety of aircraft, propellor and jet-driven, but nothing had prepared me for this: it was rather like a catapult launch indecently prolonged. Nozzles down and things become more normal – if hanging in the air at a ridiculously low airspeed can be called normal – but that, of course, was what was expected and desired. We did two launches at different weights and speeds, both well above VTO weight. In between I was allowed to fly the aeroplane about in gentle manoeuvres when it seemed much like any other aeroplane

Figure 18 Trying it for myself in a two-seat Harrier

to me. Landing, though, at well above hover weights, was an eye-opener in terms of approach and touch-down speeds, which were very high and highlighted the effect of the Ski Jump on launch speeds.

For me it was a climactic experience and one to be treasured in memory.

The trials continued in all sorts of conditions and by the end of the year had reached 12 degrees exit angle and handled a 25-knot crosswind (15 knots more than the Harrier's crosswind limitation for a runway STO) with no difficulty. Night launches were equally uneventful. The team of test pilots led by John Farley demonstrated that the forecast benefits were obtained or exceeded. John was particularly effective in selling the concept with a quiet conviction that was more persuasive than any marketing hype. His 'Here, have a go yourself' approach won over the most sceptical of hardened aviators. We had visiting pilots from the US Navy, US Marine Corps, Spanish Navy, Indian Navy, the RAF and many others. One or two made mistakes that allowed the Ski Jump to demonstrate its life-saving forgiveness. Otherwise nothing went wrong. It was total success, total vindication.

23

Rewards and Retirement, 1978–9

In 1977 I was awarded an MBE and at the beginning of 1978 the British Aerospace publicity campaign opened with a flourish. John Fozard and John Farley jointly and individually began a series of lectures and presentations to a variety of bodies in the aviation world. Reports appeared in the technical press in England and abroad and even in the popular daily papers. Suddenly, and mercifully not for long, I was famous. So too was my wife. She deserved some recognition for putting up with me and the Ski Jump over the previous nine years of indifference and opposition but there was also another reason.

The Royal Navy was and probably still is notable for its parsimonious treatment of its personnel who make suggestions or technical innovations resulting in benefits to the service. From the time that the Ski Jump firmed up in my mind I knew that I had a major innovation of great value. This was, of course, very much a minority opinion for most of the time but it was always very clear to me and naturally I hoped that in the fullness of time I would get an appropriate award. Being realistic, the reward was likely to be less than the invention deserved. My colleagues agreed that such was the likely outcome and that what was needed was some means of inhibiting the official miserly inclination. The general consensus was that I should acquire a civilian partner on the assumption that the Navy would never dare to treat a civilian in such a manner. My wife was an obvious, though far from enthusiastic, candidate. She pointed out that she would not survive any technical questioning on the subject so we hatched the plot that she had christened it Ski Jump while looking over my shoulder at my drawings. Accordingly, when it seemed at last clear that the project was going ahead, I took out a provisional patent in our joint names. This was really an empty gesture because I knew that I couldn't afford to take out a full patent. In fact the whole devious business was a waste of time because as we now know the government can steal your ideas, whoever you are and whenever it cares to, and reward you or not as seems politic at the time. In my case I think the publicity generated by British Aerospace was a very significant factor.

In the event in July 1978 we were presented with a cheque for £25,000 which to me was a great deal of money and meant that, for the first time ever, I could in future contemplate the end of the financial month with equanimity.

The award was made with a great deal of publicity and featured on the TV news. The MOD claimed that the award was the highest it had ever made to an inventor and that, furthermore, it was only an interim award pending completion of a total evaluation of the benefits to the service.

The 'highest ever' claim puzzled me because the awards for radar and the jet engine, of relatively recent memory, were much greater even if inflation were ignored. When I pointed this out my attention was drawn to the small print and its context, that is that it was the highest award made by the *Ministry of Defence*, which had only recently come into existence. Previous awards had been made

Figure 19 Receiving the cheque: Iris and Doug Taylor with the Minister (Photo: Crown Copyright)

by the relevant service ministries, Navy, Army or RAF. Then there was the interim nature of the award, of which much had been made (see Appendix 3). This was short-lived. Within a few weeks I was summoned into the presence of a senior civil servant and told that it had been decided that there would be no further payment unless the Ski Jump was adopted into use by the RAF. I thought this was a little bit below the belt but I was not unduly concerned because the enthusiastic response of RAF Harrier pilots to the Ski Jump experience made me confident that the RAF would use it. In due course I was to discover that Harrier pilots had little influence on RAF politics.

The £25,000 award was a very useful sum of money to me but it clearly was not a lifestyle-changing sum, nor would I claim that it necessarily should be. However, I said then, and believe now, that it does not compare well with awards for comparable achievements in other professions especially when the savings to the taxpayer are taken into account (see Appendices 4 and 5, the letters from John Fozard and John Farley, respectively, to the Ministry of Defence's Committee on Awards to Inventors).

So how does one assess the value, and hence the appropriate award, for a contribution like the Ski Jump? In the commercial world the problem solves itself. If the invention is successful and protected by patent a fortune is made. If I could have charged a modest £10, say, per launch my fortune would be made (well over £500,000 by now). But I was a serving officer and my customer held a national monopoly. We are forced to use comparisons to get an idea of worth. Consider the following.

Frank Whittle: £100,000 in 1948. Whittle was paid to undertake his studies and, rather surprisingly as a serving officer, he was allowed to form Power Jets Ltd. He was developing a well-known concept in parallel with foreign competitors. One of these, Hans Von Ohain, flew his engine two years before Whittle but had the misfortune to be on the losing side in the war. This is not to belittle Whittle's dedication and achievements but these are the facts – as is the fact that all modern aircraft fly on derivatives of Von Ohain's engines.

Robert Watson-Watt: £52,000 in 1957. Watson-Watt was also doing the job he was paid to do in finding applications for radio phenomena that led, eventually, to radar. He was working with known concepts, as were the Americans and the Germans.

John Randall, James Sayers and Harry Boot: £36,000 in 1957. Randall, Sayers and Boot were being paid to work on radar developments and they came up with the cavity magnetron, which arguably was the single most significant development in radar. Again, theirs was salaried, directed and commissioned research.

Douglas Taylor: £25,000 in 1978. Working alone and unpaid in his spare time without the support and encouragement of the intended beneficiaries. The Ski Jump was not a development of any known concept and no one else was considering any similar ideas.

Radar and the jet engine were world-changing technologies and the Ski Jump is not in the same class but in its own sphere it can lay claim to similar levels of impact.

More appropriate comparisons can be made with the steam catapult, the angled deck and the mirror landing sight for aircraft carriers, all very notable contributions to naval operations.

In the case of the steam catapult, this was an adaptation of the catapult used by the Germans to launch flying bombs against England. No invention was involved; indeed the concept originated in the 18th century or earlier.

The angled deck and the mirror landing sight were original inventions by serving officers in response to generally accepted needs. They were very notable improvements to the safety of naval aviation and to the reduction of the cost of such operations.

The Ski Jump was invented by me in response to a need that I alone foresaw and that was disputed both by the Royal Navy and, initially, by the designers of the aircraft. No government support was given initially for the feasibility study until this had been established by my own efforts. The invention provides very large aircraft performance improvements and flight safety enhancements together with very considerable cost savings. The elimination of the need for expensive, complex flight deck machinery and its associated operating and maintenance personnel is an additional operating and financial advantage. Assuming that any award would be based on a combination of the impact of the invention upon the state of the art and of the inventor's originality and personal input, I believed that the Ski Jump ranked somewhere between the angled deck and the mirror sight on one hand and radar and the jet engine on the other. As ever, John Farley had words to summarise it: 'The Ski Jump really was the best example of a total win-win aviation idea that I have ever come across.'

To put the affair into some sort of relative context, my £25,000 would not have been sufficient to buy one set of Buccaneer drop tanks.

Meanwhile, the trials went on at Bedford and reached an exit angle of 20 degrees, which was the limit to which the trials ramp could be adjusted. It was probably also the practical limit for the Harrier without modification and for retrospective fitting to existing ships. I had attended most of the trials despite my unofficial status and felt that they had been conducted in an exemplary manner. As a result of all the publicity, various people had reached similar conclusions to me that big ships were not essential for the effective operation of V/STOL aircraft. Vosper Thorneycroft and Vickers Shipbuilders in particular produced designs for smaller ships such as the 'Harrier Carrier' and suggested various possible adaptations of merchant ships. The Navy let these efforts pass them by without comment but let it be known that a seven-degree Ski Jump would be fitted to HMS Invincible. The ramp could not be higher, they said, because it would interfere with the firing arcs of the Sea Dart missile launcher. In my opinion this was in fact DG Ships' final piece of obstruction and I protested at it. The missile launcher could be raised at modest cost sufficiently to clear a ramp of 12 degrees exit angle and I predicted that this in fact was what they would need to do eventually. But I protested in vain and as a result the taxpayer suffered two further lots of unnecessary expense. HMS Invincible had the seven-degree Ski Jump fitted, which was later removed and replaced by one of 12 degrees; HMS Illustrious began life with a nine-degree Ski Jump, later removed and replaced with one of 12 degrees; and HMS Ark Royal had a 12-degree ramp from the start. Having had Ski Jumps forced upon him, DG Ships made maximum use of the event. Invincible was late in building and millions of pounds over budget, which is situation normal for MOD procurements and invariably leads to all-round recrimination. But here, in the shape of the Ski Jump, was the ideal scapegoat for the mounting costs and the delays. It was an opportunity not to be missed and it wasn't.

In the context of DG Ships' initial assertion that the Ski Jump could not, practically, be built to the required accuracy there now came a most interesting development – a piece of standard military equipment called the medium girder bridge. This was normally used by the Royal Engineers, as the name suggests, for bridging rivers. It was capable of carrying tanks and heavy vehicles and an ingenious sapper had discovered that by lifting one end of a medium girder bridge its segments fell into a natural arc to make a Ski Jump, only needing supports placed underneath to become fully functional in launching a Harrier. The Royal Engineers built one at Farnborough for the Display in less than 24 hours. It was the highlight of the Display, a windfall that had not been anticipated, an instant Harrier Ski Jump using existing equipment at no extra cost to the taxpayer and applicable to ships or on land. It was a brilliant example of lateral thinking for which I suspect the thinker got little recognition. For their own devious reasons of inter-service politics both the RAF top brass and the Admirals ignored it.

My big project was now generally accepted as an unqualified success and that was very gratifying. I had congratulations and compliments from all sides and the few weeks of fame were an interesting and in many ways an eye-opening experience. We were not sorry, however, that the fame quickly faded; anonymity is a more comfortable situation.

The opposition had naturally become very quiet now but there was beginning to become an insidious process of belittlement, mostly nothing more than an irritant but occasionally a source of anger. It mostly took the form of, 'What's all the fuss about? It's only a simple curved ramp.' The other main form was, 'Well, it's obvious, isn't it?' The point at issue appears to be the Ski Jump's simplicity, which is seen as a drawback rather than a desirable and not easily attainable objective. All too often, in my experience, complexity is admired, especially by academics, who, if they feel there is insufficient of it, will boost it by the use of obscure terminology. The Royal Aeronautical Society awarded me their Bronze Medal, which would have been a source of pride to me had it not

Figure 20 The medium girder bridge as Ski Jump (Photo: British Aerospace)

been for the presenter telling me that it was a 'controversial' decision in view of the obviousness of the Ski Jump. That did make me angry, so much so that I came close to walking out of the proceedings. It had been not so obvious when many members of the RAeS were proclaiming that it could not work, that it was the same as perpetual motion, there could be no gain or at best only a marginal gain. It was quite unnecessarily grudging.

Now my time in the Royal Navy was coming to an end and after 34 years I was going out on a high note. I had sometimes wondered how I would feel after so long with one organisation and in the event I felt no cause for either celebration or regret, more a kind of relief that a burden had been lifted. I left quietly on 5 May 1979, my 50th birthday.

That same weekend the *Sunday Telegraph* ran an article about me and my family in its colour supplement. I had known it was coming and had been rather dreading it after some of the inaccuracies and misunderstandings in previous reporting of the Ski Jump in the media but my fears were groundless. The article was sympathetic and, most importantly, accurate. I can only say that its author, Andro Linklater, did us proud.

I now had to find myself a job and it has to be admitted that I had had no firm ideas about what I wanted to do. At first I rather wallowed in my new sensation of freedom. I was very fortunate in that as a result of all the publicity I was not short of job offers, although I was soon to find that among the genuine ones were some very peculiar enterprises and people. In the event I joined the then Marconi Avionics and began what turned out to be an extremely interesting and varied 16 years with few if any dull moments.

24

Aftermath, 1979–82

My connections with Harriers and Ski Jumps now no longer existed but from time to time I was reminded of them. The Royal Aeronautical Society awarded me their Bronze Medal, with some reservations, as noted earlier. The Institution of Mechanical Engineers gave me the James Clayton Memorial Prize (Frank Whittle was an earlier recipient). What I most appreciated was the James Martin Gold Medal from the Guild of Air Pilots and Navigators. The really nice thing about that occasion was that John Farley was also present to receive an equivalent award.

Quite frequently there were news items in the technical press (*Flight, Aviation Week* and others) concerning the Ski Jump and for a while it looked as if the US Navy might be interested on behalf of the US Marine Corps. They had been enthusiastic during the trials at Bedford and had had two Ski Jumps built at the Naval Air Test Center at Patuxent River, Maryland. So far as I know, their results for the Harrier would have matched ours but of particular interest to me was the fact that they also tried other, non-V/STOL types of aircraft on the Ski Jump. During my own early investigations I had matched various conventional aircraft for which I had information, such as the Sea Venom, Hunter and Buccaneer, against the Ski Jump on paper. My results indicated that, pro rata, the performance gains for a conventional aircraft would be similar to those of the Harrier, that is, that a 12-degree Ski Jump would yield a two-thirds reduction in take-off distance. The very limited published information on the Patuxent River trials suggests that, in practice, these results are achieved. In most cases this means that these aircraft could take off from a full-sized aircraft carrier equipped with a 12-degree Ski Jump without the need of a steam catapult. They would still need arresting gear for the subsequent landing on the ship and this is the approach that the Russian Navy took using navalised Mig and Shukoi aircraft.

You still need big ships with this approach but the elimination of complex, vulnerable and manpower-intensive catapults is obviously considered to be well worth while. Apart from the Russians, a number of other navies are now using Ski Jumps, though without any acknowledgement either to me or to the Royal Navy.

Within a very few years the Ski Jump settled down as a part of the naval aviation scenery. It became, for the Royal Navy at least, standard operating equipment and unlikely, I thought, to make any more headlines – but I was wrong about that.

26

Operation Corporate, 1982

In early 1982 it seemed that the long-running assault by the RAF on the Fleet Air Arm had been successful and the battle was over. The RAF, skilled politicians that they are, had persuaded the government that the country could do without the Navy's aircraft and that the Invincible-class ships and their Sea Harriers should be scrapped. The then Secretary of State for Defence – an accountant or banker – unsurprisingly took the opportunity to make the appearance of saving money and agreed. Henceforth, he said, the nation's air defence needs would be safe in the hands of the RAF whenever and wherever they arose. They arose sooner than expected and, from the RAF's point of view, in a most inconvenient place.

In spring 1982 the Argentines invaded and occupied the Falkland Islands. They did this with no great difficulty because the government considered that a platoon of Royal Marines would be sufficient to defend the Islands. In the face of a whole army, a platoon, even of Royal Marines, proved to be inadequate. The question then was, should we, could we, reverse this humiliation by taking the Islands back or should we swallow the last of our national pride and accept the situation? There were plenty of those, especially on the political left, who were in favour of the latter course but the Prime Minister, Margaret Thatcher, was not a woman to turn the other cheek in these circumstances. She summoned her Defence Staff to decide on the next move. The details of the ensuing conversations have never been released but a story goes that the conversation went along the following lines.

PM: 'Can we take the Islands back?'
Army: 'Certainly, if the Navy can put us ashore.'
RAF: 'Ah, but the Islands are out of range of air cover, so it might be as well to accept the situation.'

Navy: (1st Sea Lord, seeing his chance): 'The Navy will put the Army ashore *and* provide the air cover.'

So it turned out. Very shortly the task force sailed to the applause and with the support of the mass of the British people, whose mood the PM had instinctively understood. Operation Corporate recovered the Falkland Islands at no small cost in blood and treasure. It was a real war, albeit a small one, and not a paper exercise. It exposed the deficiencies in British defence that had been caused by RAF doctrinaire mythology and the indifference of politicians of all parties. Because of its inherent limitations the RAF could have no major part in the operation and the Fleet Air Arm proved its worth. The carriers were saved, for the time being at least.

It is a nice little irony that the Argentines saved our Navy's carriers and the carriers' Ski Jumps made it possible for our carriers to contribute to Argentina's defeat. The war was won by human courage and sacrifice aided by technology. It can be argued, though, that the technical heroes of this war were the Sea Harrier, its American Sidewinder missile and the Ski Jump. The inclusion of the latter reflects a wider body of opinion than that of its proud inventor and is based on the following facts.

The South Atlantic around the Falkland Islands can generate some of the roughest seas and biggest swells in the world, causing ship motion that would preclude full-load take-off from a level deck. The Argentine aircraft carrier could not operate effectively in those conditions and turned for home very early in the conflict. By contrast the Ski Jumps in Invincible and Hermes allowed their Sea Harriers to take off with effective payloads in all sea conditions by day and night during the weeks of the operation. In addition, the ability to carry a full fuel load from a Ski Jump launch enabled the Sea Harriers to maintain a combat air patrol over the Islands while their parent ships stood off beyond the range of the shore-based Argentine aircraft and their Exocet missiles. This was vital because, as the PM herself conceded, if the carriers themselves had been sunk or disabled the operation would have had to be abandoned.

The Ski Jump had proved itself to be well worth its very modest costs.

26

And Afterwards?

Part of the fame that came my way as a result of all the earlier Ski Jump publicity included a Royal Navy recruiting advertisement with me as its central feature. Never in my wildest dreams did it ever occur to me that one day I should be publicised as a role model for recruitment to the service that had, when I was a cadet, midshipman and sub-lieutenant, regularly brought to my attention my manifold deficiencies. It is certainly fair to say that this feeling applied also to those members of the Royal Navy hierarchy who were aware of my existence, in particular my term officers at Dartmouth, in the training cruiser and at the Naval Engineering College. I do hope they all saw the advertisement. There it was and there is no denying that it gave me a great deal of wry pleasure every time it appeared. However, it went on and on. For years. Long after I left the Navy this splendid advertisement proclaimed that I was in the Ministry of Defence working on a Sea King replacement. My work colleagues at Marconi Avionics made a point of asking for an update on the situation with the Sea King replacement and could I perhaps spare a moment from it? It was fun but it had to end and so I wrote to the MOD asking for it to be discontinued. Which it was.

I sometimes wondered when the RAF would get around to using the Ski Jump so that I could claim the second part of my 'interim reward' because all the RAF Harrier pilots who had tried the Bedford installation had been instant and enthusiastic converts to the idea. They realised that they could now operate at maximum payloads from concealed sites that could be prepared in a matter of hours using the medium girder bridge. I had no doubt in my mind that it was only a matter of time before the facts filtered up to the Air Marshals and there would be another satisfied customer. I was wrong, of course. Air Marshals are perhaps the most hidebound of all the military bureaucracy and although the facts did, in time, filter up the adoption of the Ski Jump was vetoed. The reason given was that it would remove the Harrier force from a 'RAF-specific' infrastructure and would, in consequence, be contrary to the doctrine of indivisibility of air power.

There can be very little doubt that there were grounds for their misgivings because it would have shown to the merest layman that the most appropriate people to operate those Harriers are soldiers.

So there it was. There would be no further instalment of my 'interim award' and there was no point in getting particularly indignant about it. As the old saying goes, 'If you can't take a joke you shouldn't have joined.'

However, circumstances can change in the most unexpected ways and when least anticipated. In the year 2000 the MOD announced the formation of Joint Force 2000. Under this arrangement the Fleet Air Arm and the RAF would amalgamate their Harriers at a single headquarters station under the command of an Admiral. I regarded this development with the deepest suspicion: 'Beware of Greeks bearing gifts' was the phrase that came instantly to mind. Only a little later came the realisation that the terms of the arrangement meant that henceforth the RAF would be using the Ski Jump as a matter of normal operations. I wasted no time but sent off a letter to the Secretary of State for Defence claiming the rest of the award. My expectations were not high and it must be admitted that my motive was as much mischief as confidence that justice would be done.

The outcome was as I had expected, although I had wondered what the argument would be. It was that although it might happen that RAF aircraft might use the Ski Jump, this was in the interests of maximising usage of existing resources and did not amount to any additional use. Time would also show that my forebodings about Greeks and their gifts were well founded.

Then another wheel turned full circle and the Navy's star was suddenly in the ascendant, to such an extent that the government gave approval for two new aircraft carriers to replace the Invincible-class ships. The new ships are to be of 55,000 tons displacement, the biggest ever built for the Royal Navy, although only just big enough to operate modern, heavy conventional jet aircraft. The Americans and the Russians appear to think that

70,000 tons is a better starting point. Nonetheless the Navy has been granted its deeply desired CVA 01, brought up to date and doubled. The circumstances of this apparently desirable outcome were, in hindsight, more than a bit suspicious. The first oddity was the sudden scrapping of all the Navy's Sea Harriers, which were now part of Joint Force Harrier. The deed was done swiftly with minimal publicity and, incredibly, with no protest whatsoever from the Navy which, once again, was left with no airborne defence capability. One cannot avoid the suspicion that a less than honourable deal was done behind the scenes: in exchange for the Sea Harriers today, new ships – big ones – tomorrow. And with the new ships, the very latest STOVL aircraft – the F35 Joint Strike Fighter. Accordingly the ships were designed to be fitted with Ski Jumps but not with arresting gear.

I have to admit that I thought that the ships would never be built, that the government would find some excuse to cancel them and that this would not necessarily be a bad thing, especially if it forced the Navy and the MOD seriously to consider smaller carriers and a derivative of the Harrier instead of the F35. For the F35 is a hideously complex and challenging concept, combining as it does supersonic performance with stealth and a vertical take-off and landing capability. I have little doubt that the United States can deploy the engineering talent to achieve the aim eventually but it may well be a triumph of engineering over common sense and affordable only by the US Navy. For all its complexity, I could see no evidence of originality in the F35 design. It bore all the signs of design by a committee who, as an afterthought, added a crude kind of vertical landing capability. It appeared to owe nothing to the extensive operational experience with the Harriers and in particular lacked the capacity to vector its thrust virtually instantly. Transition to and from jet-borne flight would be done under electronic control, which would certainly ease the piloting task but would mean that short take-off and Ski Jump launches would be dependent entirely on the serviceability of these complex systems.

I was wrong again. The ships were not cancelled, contracts were signed, metal was cut and building was started. There were rumours of problems with the STOVL version of the F35 but everything was going ahead. Then came a new government and a Defence

Review with the stated aim of making drastic cuts. This time it seemed the new ships would be cancelled after all, expensively, in view of the start of construction, but understandable as an outcome of the deplorable indifference of politicians to defence matters and the political machinations of the RAF against the Fleet Air Arm. In the event what might only have been tragic became farcical as well.

The new ships were not cancelled. Instead HMS Ark Royal was decommissioned immediately and all the remaining Harriers of the Joint Force were taken out of service for scrapping, again with immediate effect. The effect of this was to ensure that there would be no British aircraft capable of operating from ships. The Navy was left in the ludicrous situation of having two new carriers under construction but no aircraft to put in them when the ships came into service. Strange to relate there was, once again, not a word of protest from the Naval Staff against such a grotesque decision.

There were, naturally, plenty of protests from other quarters, including me. I insisted on seeing my MP and bent her ears on the subject. She took it well but I was well aware that as a backbench MP there was little she could do. In any case the deed was done. The British Navy had become a laughing stock and that could not be undone. The population at large had neither interest in nor knowledge of the Fleet Air Arm but even so there was sufficient protest, mostly from retired officers, to provoke a response from the Secretary of State. He assured the nation that an adequate defence capability remained and that should some emergency arise our allies, the French, would provide naval aircraft to operate from our bereft ships. This took the show beyond farce to the far reaches of ridicule and must have generated a Gallic giggle across the Channel because the new ships had been designed without the arresting gear that is an essential requirement for the French aircraft. Our hero considered this to be only a minor impediment and let it be known that the missing equipment would be provided. Perhaps that is the essence of transparent government: that you make it clear that you do not understand the problem.

The Naval Staff do not have that excuse and, sadly, they appear to have connived at this and at the next move, which was to assure those few still following the proceedings that all was well because at least one of the new ships would be modified, in build, to be equipped with both catapults and arresting gear. Just like that. The catapults could not, of course, be of the tried and tested steam variety in a ship having no steam generating plant. Instead they would be entirely new electro-magnetic devices currently under development for the US Navy. It has been said that those who ignore history are doomed to repeat it. Much the same applies to those who, from ignorance or wilful obtuseness, refuse to learn from experience. The situation was both sad and ludicrous and particularly humiliating for those of us who had devoted much of our working lives to naval aviation. Surely, I thought, they can't be serious; someone will point out the folly and its consequences. No one did, at least none that I noticed, so once again I allowed myself to indulge in a gesture that I knew to be futile. I wrote to the MOD pointing out the engineering consequences and predicting that a great deal of money would be spent in vain before the technical facts forced an expensive U-turn. Eventually I received a bland and patronising reply that ignored the concerns that I had raised. Well, at least they had been told. And of course it happened just as I had predicted after much wasted time and money. I claim no virtue in this prediction; anyone with any reasonable technical knowledge of the subject (and there must be dozens of us) would have said the same.

At the time of writing, construction of the two new Queen Elizabeth class ships is well advanced and the STOVL versions of the F35 aircraft have been ordered and are doing their trials. The expense of these aeroplanes is such that we can only have a few and these must be shared between the Navy and the RAF. It seems that 100 years of aviation history and bitter experience are to be ignored and the error of 1 April 1918 repeated.

For myself, I have been sharply reminded of the Law of Unintended Consequences. I set out to find a way of operating high-performance aircraft from small ships and to a large degree succeeded in doing so. The consequence has been that my invention now appears only in the largest ships that the Royal Navy has ever had.

Winding Up

My original objective was to find a way of operating V/STOL aircraft and in particular the Harrier from frigate-sized ships and I cannot claim, beyond argument, to have succeeded because it has not yet happened and the probability now is that it never will. The attributes of the Ski Jump have been turned in a different direction to suit the circumstances (and the prejudices) of the various users. The US Navy, having tried it, has decided not to use it – probably because they don't need it, having a good many aircraft carriers in the over 70,000 tons category. I may be wrong but I can't avoid the suspicion that the fact that it makes possible the use of much smaller ships has no more appeal to the American admirals than it had, originally, for ours. The Royal Navy used it because, politically, they had no choice and they are using it in the new carriers because it avoids dependence on vulnerable and expensive launch and recovery machinery, to say nothing of the manpower implications.

Those who have the means and the freedom of choice but no inherited historical biases, such as the Russians and the Chinese,

Figure 21 The proud inventor

have chosen the Ski Jump and have chosen to use it with conventional aircraft. There are a few interesting years ahead in this area of naval aviation before the unmanned aircraft become dominant.

So I didn't succeed in getting V/STOL aircraft in frigate-sized ships but I believe that I can fairly claim to have shown how it could have been done, given the will.

That ends my part in my Ski Jump story but I feel sure that we are not yet at the end of the Runway in the Sky.

Appendices

Appendix 1

The slow launch incident: Buccaneer near failure

To launch a 20-ton aircraft from zero to 120 knots speed requires a great deal of energy applied over a very short period (about two seconds). Since the energy needed is more than is available on a continuous basis, it has to be stored for release when required. In aircraft carriers the energy source was steam and it was stored in a huge steel flask called a steam accumulator. In Victorious it was (from memory) about 20 feet long and 10 feet in diameter and was called a wet steam accumulator, because it also contained hot water. The layout is shown in Figure 22.

Steam is admitted to the accumulator via the charging valve, which is operated from the catapult control position on the flight deck. The steam passes along the horizontal pipe, which is blanked off at the end so that the steam is forced to go down the vertical pipes whence it bubbles through the water before arriving in the steam space above. In the process the steam heats the water and charging is continued until the steam pressure in the accumulator reaches the required pressure, which varies according to the type and weight of the aeroplane to be launched. At this point the steam and water are at the same pressure and temperature and are said to be in equilibrium.

To launch the aeroplane the launch valve is opened so that steam leaves the accumulator and passes to the catapult. This results in a drop in the steam pressure so that steam and water are no longer in equilibrium and the water therefore 'boils' violently, creating more steam, which reduces the steam pressure drop, thus maximising the energy applied to the launch.

The cause of the near fatal Buccaneer incident was the blanking plate at the end of the horizontal steam pipe. This was specified to be attached by means of a 360-degree peripheral grade A weld. In fact it was held by four little tack welds – a disgraceful failure of workmanship and supervision.

Remarkably those puny tack welds held through a good many launches but inevitably failed during charging and fell to the bottom of the accumulator.

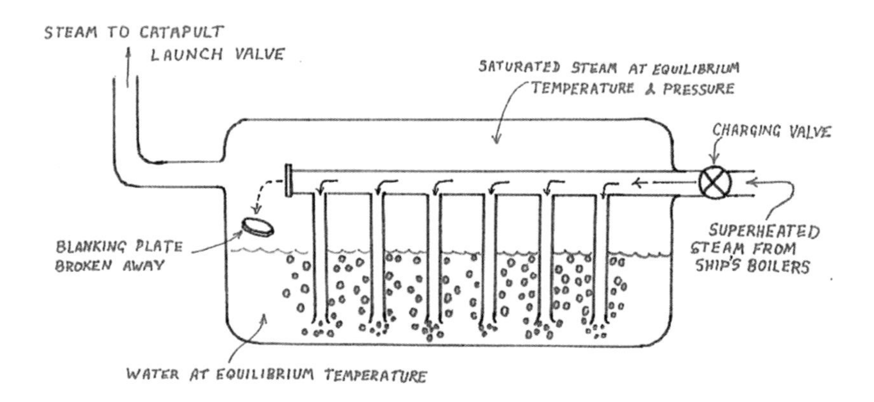

Source of a near disaster – the wet steam accumulator

Now the steam could pass straight through the horizontal pipe and out into the steam space without bubbling through the water, which therefore remained relatively unheated. When the steam pressure reached the required value and the Buccaneer was launched the steam pressure dropped but the water was relatively cool and so did not flash off into more steam to reinforce the launch energy.

Aircraft are always launched with a margin of safety in terms of speed but this failure had eaten up all the margin. Disaster was averted only by flying skill, ground effect and possibly a timely increase in wind speed.

Appendix 2

Signal about the passage through Lombok Strait

FROM. C.T.G. 320.1. ROUTINE.

TO. T.U. 320.1.1. RESTRICTED.
T.U. 320.1.2.

171400Z.

TOMORROW WE START OFF THROUGH THE LOMBOK STRAITS AND THE JAVA SEA TO SINGAPORE.

2. , WE SHALL NEED COOL HEADS AS WE MAY WELL BE SUBJECTED TO PROVOCATIVE ACTION AT CLOSE QUARTERS BY INDONESIAN SHIPS AND AIRCRAFT. THIS WE SHALL IGNORE.

3. BUT IF ANYONE OPENS FIRE ON US I SHALL ORDER YOU TO RETALIATE AND WE WILL GIVE MUCH BETTER THAN WE GET.

4. GOOD FORTUNE.

D.T. G....171400Z. SEP. '64.

RESTRICTED.

REL:- S.O.O.
DIST:- FULL - STAFF 2 FOR TX.
OUT NR 24/17 PL . LIGHT P.G. TOD I/T 17/9/64.
GREEN. EGAN.

Appendix 3

Letter re 'interim payment'

PROCUREMENT EXECUTIVE, MINISTRY OF DEFENCE

St Georges Court, 14 New Oxford Street, London WC1A 1EJ

Telephone (Direct Dialling) 01-632 3600

(Switchboard) 01-632 3333

JY/3500/0610 3 July 1978

Lieutenant Commander D R Taylor, MBE, RN

Sir,

I am directed to inform you that it has been decided to make an interim ex-gratia award of £25,000 from public funds to yourself and Mrs Taylor jointly in respect of the Ski Jump launching device for V/STOL aircraft.

This award will be reviewed in due course when the full benefit to the Crown has been more clearly established.

The enclosed form of acceptance should be completed and returned to the address indicated so that payment can be arranged.

I am, Sir
Your obedient Servant,

E. A. Hedan

Appendix 4

John Fozard's letter to the Committee on Awards to Inventors

British Aerospace
AIRCRAFT GROUP
KINGSTON – BROUGH DIVISION
RICHMOND ROAD KINGSTON-UPON-THAMES SURREY KT2 5QS

JWF/LA

3rd January, 1978

R.F. Smith, Esq.,
Secretary
Committee on Awards to Inventors,
Ministry of Defence (PE),
Pats 3B, Room 2104,
Empress State Building,
Lillie Road,
LONDON, SW6 1TR.

Dear Mr. Smith,

SKI-JUMP. AWARD TO LT. CDR. D.R. TAYLOR, R.N

Thank you for your letter of December 5th. I hope this reply will be in time for the January meeting of your Committee. As will become clear, I believe that Lt. Cdr. Taylor has a substantial case for an award in respect of future benefits to the Crown arising from his origination of the Ski-Jump launch concept. This letter puts forward the basis of my belief.

The Ski-Jump trials being conducted at RAE Bedford have now reached a launch angle of 12 deg. The earlier testing at 6 deg. (August 1977) and 9 deg. (Oct/Nov 1977) fully confirmed the predictions made by HSA during the past several years of analytical study. The 12 deg. trials to date (their completion is not due until mid-January, 1978, with 15 deg. launches scheduled for late Feb/early March 1978) have also confirmed the accuracy of our predictions and demonstrated that the technique is overwhelmingly Good News.

Let me make it quite clear from the start what are the benefits of the Ski-Jump launch technique from ships at sea. Bear in mind in what follows that the technique applies substantially only to vectored thrust V/STOL jet aircraft exemplified by the Harrier, Sea Harrier, AV-8A, AV-8B; there might be an application to conventional jet naval aircraft, but the potential gains are small and only the U.S. Navy would be in a position to exploit this aspect since, by the mid-1980's, the USN will be the only service flying conventional fixed-wing jets from decks.

The direct benefits of a Ski-Jump launch are

1. Increased launch PERFORMANCE

and 2. Increased launch SAFETY

Indirect benefits which follow from these, realisable by ships carrying the Ski-Jump, are:-

3. Greater endurance at sea because of the lower average ship speeds needed to launch aircraft.

and 4. Smaller ships becoming capable of providing a given level of Harrier sortie capability at sea.

TELEPHONE 01-546 7741 TELEGRAMS BRITAIR KINGSTON-UPON-THAMES TELEX 23726

2.

All the above benefits have clearly been demonstrated by - or will follow from - the analyses and test flying already accomplished. I have no reason to suspect that the future work at RAE, taking us to the full 20 deg. launches by mid 1978, will reveal any lesser standard of achievement, in terms of the still larger potential gains. HMS Invincible (the first of the new class of Command Cruisers for the RN) will be fitted with a 7 deg. Ski-Jump when she goes to sea in 1979. The 7 deg. angle results from the RN's wish to avoid restrictions on the (unmodified) Sea Dart Launcher at the front of the flight deck, and the fact that when the decision to fit a ramp was made (in September, 1977) we had confirmed only the 6 deg. launch by test. I believe that the second and third ships of this class could emerge in the 1980's with ramps providing a launch angle of 12 to 20 deg., depending on how much the RN are prepared to modify these later ships.

. The PERFORMANCE gain for a 7 deg. Ski-Jump can be expressed as 1500 lb more payload compared with a corresponding flat deck launch. If we assume that later ships will have a steeper ramp, a performance gain of at least 2000 lb. in launch payload can be readily anticipated for the Harrier at sea in the 1980's. How do we cost this benefit?

This increase in payload (fuel plus weapons) could be provided by augmented engine thrust when launching from a flat deck. A gain of 2000 lb. in payload would need an increase approaching 2500 lb. in engine test bed thrust. Fortunately we have a recent quantification of what a comparable amount of extra thrust would cost to develop. A proposal to increase the Pegasus thrust by 1500 lb. was recently made to the U.S Navy as part of the development study of the AV-8B Advanced Harrier. The programme was estimated to cost some $ 45M. Since thrust increase is probably not linear with cost in such a development, a 2500 lb. boost would likely cost of order $65M i.e. about £36M. This amount of thrust increase from the Pegasus would need a new intake for the Sea Harrier (as indeed is featured in the AV-8B) and my guess is that this aircraft redesign would cost of order £2M plus proving/test flying of further order £2M.

Each Pegasus at the new thrust rating would be more expensive (cost approx proportional to thrust) so that for the 36 Pegasus engines on order for 24 RN Harriers, at a basic present cost of some £0.5M per engine, the production cost of the increased thrust would be $36 \times £0.5M \times \frac{2500}{21500}$ (21,500 lb being the present thrust rating). This amounts to some £2M. I assume here that the new intake could be built in production for the same cost as the existing version. Retrofit would be very expensive and unacceptably time-consuming.

A rough cost equivalent to the Ski-Jump gain in performance terms, therefore, would be 36 + 2 +2 + 2 = £42M, using today's engine technology.

If we tried to develop a 2000 lb. launch payload gain via improved aircraft wing aerodynamics, it would prove cheaper (perhaps only half the cost of the engine development route) but it would take longer and heavily compromise the other performance virtues of the Harrier - mission capability and, most importantly, vertical recovery to the deck. So I believe my cost equivalence based on engine performance is as fair as possible in the circumstances.

3.

£42M is thus the best quick approximation for the cost equivalent of the performance gain made possible by the Ski-Jump. This figure represents a "saving" to the Crown for the operational performance advantages arising from the future use of the Ski-Jump by the Royal Navy during Sea Harrier operations afloat.

The SAFETY asset of the Ski-Jump can also be quantified in terms of value accruing to the Crown. An initially upward trajectory in launch from a pitching deck removes almost entirely the chance of a sea strike due to pilot misjudgement or other unforeseen combination of the many physical factors involved. Here we must exclude such foreseeable, but unallowable, factors such as engine failure; although a nozzle system failure at launch could be survivable off a Ski-Jump, unlike from a flat deck, as pointed out in my Mitchell Memorial Lecture.

With 24 Sea Harriers each flying say 250 hours/year for 15 years from 1980, of which half the sorties are likely to be from deck launches at the rate of one launch per flying hour, the total of deck launches approaches 50,000. On present RN planning, over 75% of these will be STO rather than VTO launches - say 40,000 STO's from a deck.

At present with conventional catapult-launched fixed wing aircraft at sea, the loss rate at launch from all causes is of order one per 5 to 6 thousand. If Sea Harrier follows this trend then, in 40,000 STO's, there is an even chance of 7 to 8 aircraft being lost in launches from flat decks, over a period of 15 years. At a guess I would expect half of these losses to be of a nature where "things went wrong" without involving any major mechanical/electrical malfunction. Thus, removal of the sea strike risk at launch which is the real safety virtue of the Ski-Jump could (leaving all other risks unchanged) be the means of saving 2 to 4 Sea Harriers in the service life of the fleet. With a Sea Harrier valued at £4M, the cost saving to the Crown is thus of order £10M if we allow also for the considerable cost invested in the pilots who would otherwise be lost with the aircraft at launch.

I feel somewhat less qualified to estimate any Crown benefits that may accrue on the effects of the Ski-Jump on ships, but the numbers below will give an idea of the order of their importance.

The 6 deg. Ski-Jump has a launch effect corresponding to a 15 knot Wind-Over-Deck (WOD) in typical launch conditions. Steeper-angle ramps will have WOD effects equivalent to over 25 knots.∴ Thus, in atmospherically calm conditions, ship speed through the water for aircraft launch need be only half that necessary with a flat deck ship. Or, in light winds, the ship need not turn into wind for aircraft launch because of the WOD advantage given by a 12 + deg. Ski-Jump. Since ship propulsive power varies approximately as the cube of water speed, halving the water speed cuts the power requirements by $\frac{7}{8}$. For a ship such as Invincible this reduces shaft power (for a reduction from say 28 knots to 14 knots) by over 80000 HP. The corresponding ship fuel consumption will be reduced by some 20 tons/hr. If we assume 10 launch details per flying day at sea, each launch cycle taking say 15 min, the economy of fuel due to the Ski-Jump will be 20 tons/hr x 0.25 hr x 10 = 50 tons per day. We might note in passing that this is the same order as the fuel consumed in a 24 hour period at normal (12/16 knot) cruise speed, so that an extension of range/endurance at sea of upwards of 25% between replenishments is feasible compared with the non-Ski-Jump-equipped Harrier ship. There must be some value we can assign to this in terms of the cost of less frequent rendezvous with RFA vessels. This will not be quantified here, however.

4.

If each Invincible class ship spends on average 150 days at sea for 20 years, and flying takes place on 100 of those days, and on 30% of those flying-at-sea days the natural wind is light enough so that the gain from the Ski-Jump effective WOD is realisable in terms of ship speed through the water, the fuel saved compared with flat-deck ships will be:-

30% of 100 days/yr x 50 tons/day x 3 ships for 20 yrs = 90,000 tons. At an average price of 50p per gallon over the period 1980/2000 the cost saving to the Crown will be some £15M, attributable to the effects of the Ski-Jump on the operation of Harrier-carrying ships.

Assigning a value to the smaller ship which will in the future become practicable for Harrier operations as a consequence of the Ski-Jump is much more difficult. My Mitchell Memorial lecture discusses future ships including the Vosper Thornycroft Harrier Carrier (which a 20 deg Ski-Jump would make incontestably viable). For example, the last of the 3 planned Command Cruisers for which contracts have not yet been let, could be replaced by two Harrier Carriers for about the same total cost. But that would be no procurement saving to the Crown, and how does one place a value on an extra Sea Harrier ship in the Royal Navy of the late 1980's? In the much longer term, say the 1990's, if it becomes necessary for any reason to replace an Invincible Class ship, the Ski-Jump makes practicable a 10000 ton hull as distinct from the 20000 ton Invincible size. Thus the cost of a replacement ship should be reduced by well over ⅓rd which is likely to represent a sum substantially in excess of £100M at "then" prices. But this saving to the Crown can be only notional at present.

Summing up the benefit to the Crown from the adoption of the Ski-Jump by the Royal Navy, therefore, there is a tangible gain of £25M attributable to reduced Harrier launch attrition plus the economy of ship fuel for launch over the next 15 to 20 years. In addition, the performance gain given by the Ski-Jump has a cost equivalence, in terms of conventional aeronautical technology, of order £40M. This represents a very substantial advantage to the Crown, and could not otherwise have been provided in a Sea Harrier for the RN.

The total Crown benefit arising from UK use of the Ski-Jump is therefore in excess of £65M.

Your letter asks how the Ski-Jump concept might stimulate future exports of the Sea Harrier. This is difficult to answer as, of course, there is no real cause-and-effect relation. The best that can be said is that the Ski-Jump provides the ship-based Harrier with a much better operational capability. Hence it can be seen greatly to reduce any doubts that may be in the mind of potential customers, particularly those who are constrained to deploy Harriers on smaller ships, which I fancy will be the norm in future fleet air arms.

The following lists the candidate Sea Harrier users who may be expected to order the aircraft within the next 6 to 8 years, together with a probability assessment based on our own market surveys and in some cases, direct discussion.

5.

User	Likely date	Number of Harriers	Probability of Contract
Spanish Navy	1981/2	12+	95%
Indian Navy	1982/3 1985/6	8 8	90%
Royal Aust. Navy	1981/2 1985/6	5 10+	80%
French Navy	1983/4	20	25%
US Navy	1982/3	30+	20%
Imp. Iranian Navy	1984/5	20+	10%
Italian Navy	1983/4	12	10%

TOTAL 125

TOTAL (Number x probability) 50

Thus there is a high probability of some 50 Sea Harriers being exported, with longer odds on this number being doubled within the next 8 years.

The Spanish Navy have asked us to advise on the fitting of a Ski-Jump to their existing wooden-decked ship, SNS Dedalo, from which their current Harriers ("Matador" in the Armada Espanola) already operate. They intend to build 3 new Harrier-carrying ships, based on the 1974 USN Sea Control Ship design, and these 14,000 ton flat-tops will be fitted with a Ski-Jump. The keel of the first ship is already laid at El Ferrol in N.W. Spain.

The Indian Navy cannot embody a Ski-Jump in their existing ship, INS Vikrant, as they need to operate conventional Alizé ASW/AEW aircraft with the Harriers. The requirement for catapult launch and arrested landing of the Alizé aircraft prevents a Ski-Jump being fitted either on the axial, or on the angled, decks. However the Vikrant replacement, planned for the end of the 1980's, will have a Ski-Jump.

The RAN are currently running a world-wide design competition for a replacement for HMAS Melbourne, for commission in 1985. British Shipbuilders (Vosper Thornycroft and Vickers) are bidding, and HSA are working with both ship design groups on the ship/aircraft interface, including Ski-Jump. The RAN have specified the Harrier in their design data for this future ship. If the UK win the contract to build the prototype RAN ship (with 2 further ships to be built in Australia) the UK Balance of Payments could benefit by well over £150M on the shipbuilding front. The RAN ship is made more practical (smaller,and hence less costly, for a given level of Harrier capability) by virtue of the built-in Ski-Jump, and the Ski-Jump technical expertise resides in UK. This

6.

factor will help considerably to sway the decision on contract award.

Both the French and Italian navies are proposing to build new flat-top ships for the later 1980's, without catapults or wires. The Harrier is the only jet fighter that could operate from such a deck. A Ski-Jump would offer important safety and performance gains which neither Navy could afford to ignore. The UK is unlikely to earn much money from the ship side of these ventures, however.

The US Navy, in my view, are going to be driven into a crisis — like corner in the next few years by their past neglect of V/STOL. The USN intend to develop their own V/STOL fighter. A version of the USMC's Advanced Harrier, AV-8B, is possible by 1985/6, but they really require a supersonic jump-jet, which they concede can not be developed before the 1990's. Meantime, the Russian Navy has Kiev in commission, Minsk afloat, and two more of this class building. It is not hard to foresee that the USN will need to acquire some instantly-operational V/STOL fighters by the early 1980's; and where else can they go but Kingston-upon-Thames, as did the USMC for practical V/STOL in 1969? And when the USN come to fly Harriers, they will undoubtedly apply the benefits of Ski-Jumps to their operations from Harrier-carrying ships. But again it is unlikely that the UK will earn much from technical assistance in this area to the USN.

It is sad for both the Crown and we taxpayers that Taylor's original MOD-sponsored patent was not taken out in all the overseas nations who currently fly fixed-wing at sea. It is probably too late now for this to happen, but HSA are ensuring that their Ski-Jump patent applications are being filed in all candidate Sea Harrier countries. Hindsight makes it easy now to see the consequences stemming from early Establishment doubts on the potential value of Taylor's original work. Indeed , even after HSA had subsequently developed Taylor's work in much more detail, and made known their confidence in the potential benefits, official enthusiasm seemed hard to come by. It was not until the first ramp trials had been made in August, 1977, that there was any real expression of service belief in the Ski-Jump. But perhaps all of this is in the nature of the present-day MOD processes, as I touched upon in my RAeS paper.

For the record I need to set out by belief that it was HSA Kingston's associated and enthusiastic backing in 1974 which set the Ski-Jump train rolling. Taylor gave a presentation of his thesis ideas in the Old Admiralty cinema, Whitehall, on 1st February, 1974. Without the strongly-voiced support from HSA before that audience (drawn from MOD(PE) and MOD) I am sure that the Ski-Jump idea would have been quietly buried. Too true.

From that point onwards, the banner was carried by HSA, initially on our own funding and, from late 1974 onwards, partly under MOD(PE) contract. For almost two years Kingston was the sole source of Ski-Jump expertise, in terms of aircraft performance, handling, etc., and in respect of ramp profile, undercarriage loads, etc. Latterly, RAE Bedford have developed their own expertise in performance prediction,but landing gear predictions and support on ramp geometry still resides principally at Kingston. The RAE ramp trials are a joint HSA/MOD concern but technical progress is very much an HSA responsibility.

7.

Whilst the initial concept was sparked off by Lt. Cdr. Taylor, the major ideas development since 1973 has been contributed by T.S.R. Jordan and D.C. Thorby of the Kingston Design Team and by J.F. Farley, Deputy Chief Test Pilot, to identify the principals amongst those w-o have been involved.

Thus, to attach a value to the Crown of the Ski-Jump concept, via its influence on future exports of both Sea Harriers and ships is necessarily a subjective, if not an arbitrary,process.

In its present state of development, due largely to the past 4 years of HSA inputs, I would guess that the Ski-Jump has provided a 10% better probability that the Sea Harrier will be bought by foreign navies. And a Ski-Jump on the ship to replace HMAS Melbourne will provide it with a 10% better chance of surviving the procurement processes in Canberra, because a Ski-Jump ship can be shown to provide a better Harrier capability at Sea.

A Sea Harrier at early 1980's prices will cost at least £5M. There is a well-known rule that 10 year's spares represent as much investment again, with a military aircraft, as the original purchase.

So 50 exported Sea Harriers represent a return to the UK Balance of Payments of some £500M. All of this is of benefit to the Crown. But, if you prefer a more "auditable" approach, we can take that fraction of the gross (arguably between 30% and 50%) which passes as a result of taxation (direct or indirect, Company or personal) to the Exchequer when overseas contracts of this sort are executed. Thus Crown benefits from Ski-Jump "influence effects" on Harrier exports would be, say, (40% of £500M) x 10% influence factor = £20M.

Correspondingly, if British Shipbuilding win the contract for the new RAN ship, the Crown will benefit to the further extent of (40% of £150M) x 10% influence factor, which is approx £6M.

Making some allowance for further Sea Harrier exports, the <u>total</u> benefit to the Crown via the Ski-Jump "influence factor" is thus probably of order £30M. And if we choose to attach a minimal "influence factor" of order 3% to 5% to the contribution of the Ski-Jump to aircraft and ship future exports, the Crown benefit still exceeds £10M.

Adding this to the direct benefit adduced on Page 4 gives a grand total ranging between £75M and £95M for the rewards/savings to the Crown arising in the next 15 years or so as a result of Lt. Cdr. Taylor's original Ski-Jump discovery.

I apologise for the length of this letter but I felt it necessary to spell out the assumptions and rationale behind all my calculations. Of course, as you noted in your letter, we cannot be sure for a number of years of the reality of the financial harvest to the Crown and to the UK, but it seems certain that the total will be measurable in scores of millions.

8.

It is my firm belief that the Ski-Jump launch technique, first discovered and published by Lt. Cdr. Taylor in a Harrier context, will come to be regarded, deservedly, as a famous contribution to V/STOL operation, on land as well as, most notably, at sea.

Yours sincerely

J.W. Fozard

J.W. Fozard
Executive Director & Deputy Chief
Engineer(K) Chief Designer Harrier.

Appendix 5

John Farley's letter to the Committee on Awards to Inventors

British Aerospace
AIRCRAFT GROUP
KINGSTON — BROUGH DIVISION
DUNSFOLD AERODROME GODALMING SURREY GU8 4BS

14th March, 1978.

R.F. Smith, Esq.,
Secretary,
Committee on Awards to Inventors,
Ministry of Defence (PE),
Pats 3B,
Room 2104,
Empress State Building,
Lillie Road,
LONDON SW6 1TR.

Dear Mr. Smith,

SKI-JUMP - AWARD TO LT CDR D.R. TAYLOR, RN

Thank you for your letter reference JY/3500/0610 dated 5th December 1977 on the above topic.

In discussing the actual use of the ski-jump one has to consider four broad aspects, namely performance, piloting, safety and engineering. The pilots' views on the performance and engineering implications are probably no different to anybody else's, and I am sure you have plenty of inputs from others on these matters. I will be content to just say that the performance improvements that have been demonstrated are very large and sufficient to make the Sea Harrier plus ski-jump combination a big advance over the original Sea Harrier flat deck case. This improvement in performance can be taken as an extra payload at launch, as a reduced take off run, as an increased tolerance to ship motion effects, as an increase in maximum operating ambient temperature, as a reduction in ship speed or surface wind requirements or as a mixture of several of these factors combined together. One must not generalise on performance matters because only specific numbers are correct but as a guide the STO minus VTO payload increase is of order 20%, the take off run reduction 50% and the change in allowable ship pitch motion 100%. These impressive numbers apply at the ramp angles so far tested (max 12°) with even greater benefits likely to be demonstrated at greater angles.

On the engineering side only small changes to the undercarriage, to allow it to accept the loads imposed by the ramp, are needed to extract maximum value from the ramp concept on long (500 ft +) ships with lesser need for modification on shorter flight decks. These engineering changes to the aircraft, together with the attendant need to build the ramp into the ship, are a remarkably small price to pay for the extra capability.

2/...

TELEPHONE: CRANLEIGH 2121 TELEGRAMS: BRITAIR DUNSFOLD GODALMING TELEX: 859475

- 2 -

Turning to piloting matters it is quite simply easier to take off from a 6°, 9° or 12° ramp than from a flat deck and a lot easier than from a runway. I will go into the details of why this should be later, but I wish to emphasise at the outset (and before you may get bored with my diatribe!) that the performance benefits have been achieved in parallel with a reduction in the piloting task, not at the expense of more demands on the pilot. This is in remarkable contrast to the normal aeronautical scene so far as improving take off performance is concerned. Digressing, the improvements in achieved take off performance in civil aviation demonstrated by noise abatement technique now used (more climb angle or less power used must equate to greater performance if all other factors are held constant) has been obtained solely at the cost of increased demands on piloting accuracy, extra manipulations in the cockpit and reduced total safety factors. The ski-jump story is the opposite as I shall now try to explain.

Consider first the Harrier STO from a runway. Here the Harrier pilot commences a ground run exactly as if in a normal jet fighter towards that speed at which his wing lift together with his available engine lift will exceed his weight. As he reaches this speed he selects the engine nozzles down and the Harrier leaves the ground under the combined effect of wing and jet lift. As the Harrier becomes airborne the pilot must control the attitude in all 3 axes. In roll and yaw he uses the ailerons and rudder in the normal manner to keep the aircraft wings level and flying without sideslip. In pitch he is required to raise the nose with respect to the ground roll attitude in order to achieve the particular value of wing incidence on which the climb performance is based. This task of setting the desired incidence is difficult because the Harrier is unstable in pitch when it is, as here, partially jet borne. To further complicate matters there is a large longitudinal trim change due to ground effect as the initial reflection of the jets by the ground is reduced by increasing height. (Thus causing a large change in induced downwash effects around the tailplane.) This trim change passing out of ground effect is approximately ¼ of the total tailplane travel.

When taking off a flat deck the aircraft goes from on its wheels to an immediate free air situation (with the deflected jets able to exhaust downwards without reflection by the sea into the vicinity of the tailplane). This eases the pilot's task, compared with the runway case, but the flat deck case brings with it the need to rapidly rotate the aircraft nose up after launch. Until this is done the aircraft sinks towards the sea which is not normally all that far away in the first place.

From a ramp however the Harrier is delivered into free air, in a climbing attitude and with a positive rate of climb. In short in just the condition that a pilot has to strive to attain from a runway or flat deck. This reduction in pilot workload reduces the need for skill, reduces scatter in achieved performance and is most heartwarming at night.

The safety of a ski-jump launch compared to a flat deck results from the guaranteed time of flight that passage up the ramp ensures - in just the same way that a motor cycle or stunt car driver can obtain time in the air without lift, the Harrier will fly for a while without engine or nozzle deflection. Escape requires about 2 seconds for the pilot to appreciate the problem, decide to eject and physically operate the seat. This sort of time is not achievable in the worst flat deck case. It is guaranteed again with the ski-jump.

3/...

- 3 -

In considering the value of this invention it will clearly be hard to quantify the arguments. One thought I would leave with you is that the Sea Harrier was going into service without the ramp. There is no doubt in my mind whatsoever that like that sooner or later an aircraft and a pilot would have been lost, possibly several. Now we should never lose a pilot on short take off and may not lose the aircraft. Because of the increase in launch performance we may now hit targets that could not have been reached before. How do you cost the value of that? With a lot of noughts, I would suggest.

Yours sincerely,

J. F. Farley
Chief Test Pilot

Appendix 6

DG Ships throws a spanner in the works

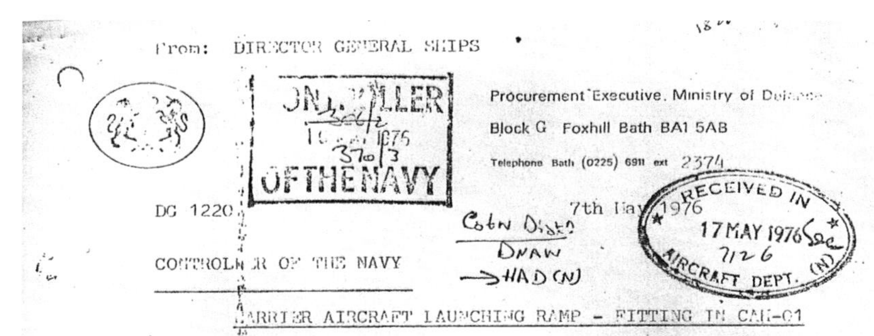

From: DIRECTOR GENERAL SHIPS

Procurement Executive, Ministry of Defence
Block G Foxhill Bath BA1 5AB
Telephone Bath (0225) 6911 ext 2374

OF THE NAVY

RECEIVED IN 17 MAY 1976 AIRCRAFT DEPT. (N)

DG 1220

7th May 1976

CONTROLLER OF THE NAVY

Cotn Distn
DNAW
→ HAD (N)

HARRIER AIRCRAFT LAUNCHING RAMP - FITTING IN CAH-01

When you visited HAD(N) you were told of the very considerable gain in aircraft performance which could be obtained by giving the forward end of the flight deck a 6° upwards incline.

2. This has been investigated in the context of CAH-01 and has been discussed with Mr Fozzard, the Harrier Chief Designer at Hawker Siddeley. It has been established that:-

a. the operational advantage offered by the mini-ramp is to allow the effects of ship pitching motion to be ignored. In the absence of pitching, the gain in take-off weight and/or reduction in required wind over the deck is relatively small;

b. the "very considerable gain in aircraft performance" would require ramp angles of 12-15° or even 20°. The higher increase would probably require modifications to the aircraft undercarriage;

c. Hawker Siddeley are lobbying for support for a development programme aimed at achieving the higher performance gains at b. above, and agree that there is no strong case for fitting a mini-ramp now in CAH-01.

3. Mr Fozzard is visiting Ship Department on 8 June to discuss the technicalities in more detail.

4. The additional direct cost of fitting a 6° ramp in CAH-01 would not be very significant, but the introduction of such a change would certainly involve delay and would also affect the launch date.

5. The Project Director has recommended against investigating and introducing such a change in CAH-01 during build and I have ruled that this recommendation be upheld.

6. For information.